General Class

FCC License Preparation
for
Element 3
General Class Theory

by
Gordon West
WB6NOA

Eighth Edition

Master Publishing, Inc.

Gordon West, WB6NOA
FCC Amateur & Commercial Radio License Preparation Study Materials

Technician Class, Element 2
Technician Class study manual
Audio Theory Course on CD
Technician Class book with
    W5YI HamStudy Software

General Class, Element 3
General Class Study Manual
Audio Theory Course on CD
General Class book with
    W5YI HamStudy Software

Extra Class, Element 4
Extra Class Study Manual
Audio Theory Course on CD
Extra Class book with
    W5YI HamStudy Software

GROL+RADAR
General Radiotelephone Operator License
Plus Ship Radar Endorsement
FCC Commercial Radio License Preparation for
Commercial Elements1, 3, and 8 Question Pools
GROL+RADAR book with
    W5YI CommStudy study software

This book was developed and published by:
Master Publishing, Inc.
Niles, Illinois

Editing by:
Pete Trotter, KB9SMG

Photograph Credit:
All photographs that do not have a source identification are either from the author,
or Master Publishing, Inc. originals.

Cartoons by:	*CD mastering by:*
Carson Haring, AC0BU	Mike Koegel, N6PTO, Studio 49

*Thanks to the following for their assistance with this book: Suzy West, N6GLF; Ed
Collins, N8NUY; AMSAT and TAPR; John Johnston, W3BE; California Rescue
ARES Net members; Question Pool Committee members; WA6TWF Super System
members; Don Arnold, W6GPS; Chip Margelli, K7JA; Amber Murphy, KD8IVE;
Walter Godoy, KI6ZVL; Tom MacKay, W6WC; and Clark Wood, K7KDF.*

Eighth Edition
   5   4   3   2   1

Table of Contents

QUESTION POOL NOMENCLATURE

The latest nomenclature changes and question pool numbering system recommended by the Volunteer Examiner Coordinator's Question Pool Committee (QPC) have been incorporated in this book. The General Class (Element 3) question pool has been rewritten at the high-school reading level. This question pool is valid from July 1, 2011 through June 30, 2015.

FCC RULES, REGULATIONS AND POLICIES

The NCVEC QPC releases revised question pools on a regular cycle, and issues deletions as necessary. The FCC releases changes to FCC rules, regulations and policies as they are implemented. This book includes the most recent information released by the FCC at the time this copy was printed.

Welcome to *General Class*!

The General Class license gives you privileges on all of the worldwide amateur radio bands for long-range skywave communications. These new bands will add to your Technician Class VHF and UHF privileges, too. With General Class, you will receive operating privileges on every ham radio band there is for voice, data, video, and CW.

As a General Class operator, you'll be able to travel the world and always stay in touch using skywave communications. Whether you sail the international waters of the South Seas, or cruise the highways of North America or Europe, you can stay in touch with your new General Class worldwide band privileges. More good news – recent international agreements allow you to operate in many foreign ports and countries without having to officially sign-up for a reciprocal license. We discuss these new agreements with Europe and South America inside this book.

As a General Class operator, you'll be able to help our hobby grow! I encourage you to join an all-General Volunteer Examination team to conduct Element 2 Technician Class written exams, and to actively recruit new ham radio operators to our hobby and service.

This new Eighth Edition of my book covers everything you need to know to pass the General Class Element 3 written examination. There no longer is a Morse code test required for any class of ham license! The Federal Communications Commission eliminated the Morse code test requirement effective February 23, 2007.

The question pool in this book is effective July 1, 2011 through June 30, 2015. The only prerequisite for your new General Class license is a valid Technician Class license or a recently-earned CSCE for Technician Element 2. And with the elimination of the Morse code 5 wpm test, this new *General Class* book will make your test preparation a breeze. We've completely reorganized the entire, new Element 3 General Class question pool to improve your learning experience and cut down on the study time required to pass the theory exam.

I am on the air daily, and I hope to talk to you soon on the worldwide General Class bands!

73!

Gordon West, WB6NOA

About This Book

My book provides you with all of the study material you need to prepare yourself to pass the Element 3 General Class examination. The precise test questions and 4 distracters are in this book. We have reorganized all of the questions and answers into common subject groups for more logical learning. This will cut down on your study time and improve your knowledge of how to operate as a General Class ham!

The Federal Communications Commission completely eliminated Morse code testing as a requirement for an amateur radio license effective February 23, 2007. I always encourage our students to know the code, so this book has a complete chapter on fun ways to begin learning Morse code once you are on the air as a new General Class operator. Chapter 5 gives you sound advice on how to learn the Morse code. Even though there is no more code test, knowing the dots and dashes will help you become a better General Class operator.

My book also provides you with valuable information you need to be an active ham on the HF worldwide bands. To help you get the most out of *General Class*, here's a look at how the book is organized:

- *Chapter 1* explains in detail the new HF operating privileges you'll earn by passing your Element 3 General Class written theory examination.

- *Chapter 2* provides a look at the history of amateur radio licensing in the U.S., and the current status of all license classes.

- *Chapter 3* contains all 456 Element 3 General Class questions, 35 of which will be on your upcoming exam. You'll notice that the questions and answers have been reorganized to follow the syllabus I use to teach General Class at my Radio School weekend sessions. It makes your learning experience easier and more meaningful.

- *Chapter 4* tells you how to find an exam site, what to do before you get to the site, what required papers to bring with you, and what to check for when you pass your exam.

- *Chapter 5* gives sound advice on how to learn Morse code at 5 wpm.

- The *Appendix* contains a list of VECs, a Glossary of important ham radio terms, and other useful information.

If you need additional study materials, I've recorded an audio course that follows this book, and W5YI also has interactive software to allow you to study for upcoming exams on your PC. In addition, I also have recorded audio courses for learning Morse code.

Need additional study materials?
Call W5YI at 800-669-9594 or go to www.w5yi.org

General Class Privileges

The General Class amateur operator/primary station license permits you to use segments of every worldwide band for long-distance voice, data, and video amateur communications. You also keep all of the privileges you now enjoy as a Technician, or grandfathered Novice or Technician-Plus operator.

The General Class license has always been considered "the big one" because of the almost unlimited skywave privileges this license provides on each of the worldwide ham bands. Although the older, grandfathered Novice and Technician-Plus licenses offer slivers of worldwide band privileges, it is the General Class license that gives you major operating "elbow room" on segments of all the worldwide bands. Whether it's day or night, summer or winter, these worldwide bands will always offer skywave communications for thousands of miles of range.

Gordo at the controls of the United Nations Station, 4U1UN, in New York.

Never in the history of amateur radio has it been easier than now to obtain the General Class license. There is no more Morse code test. The Federal Communications Commission concluded "…this change (eliminating the code test) eliminates an unnecessary burden that may discourage current amateur radio operators from advancing their skills and participating more fully in the benefits of amateur radio." Most other countries have dropped their Morse code test requirements, too.

The Element 3 theory exam has been dramatically improved in content, better reflecting General Class worldwide radio equipment and antennas. Simple math questions deal with modern, high frequency equipment. No longer will you need to memorize "far out" formulas that you would rarely use as a licensed ham radio operator. There is more emphasis on General Class operating, band plans, data modes, and high frequency simple antennas to build. Although this new question pool is more comprehensive in entire content, your upcoming exam remains at 35 multiple choice questions, 74% passing grade.

WORLDWIDE SPECTRUM

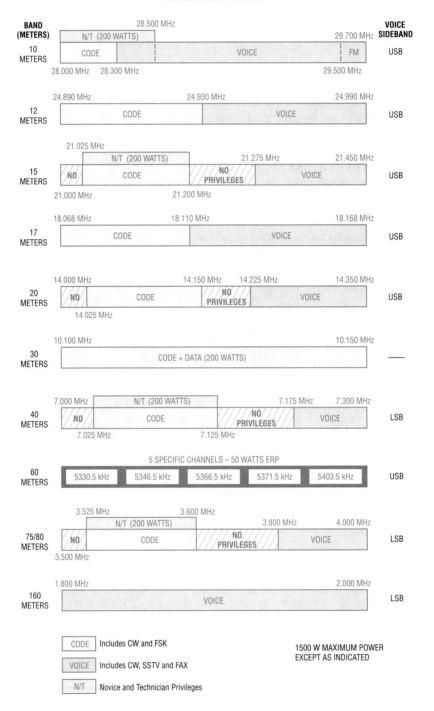

Figure 1-1. General Class HF License Privileges

GENERAL CLASS LICENSE PRIVILEGES

Figure 1-1 graphically illustrates your new General Class code, data, and voice privileges on the medium frequency (MF) bands (300 kHz-3 MHz), and high frequency (HF) bands (3 MHz-30 MHz). Code and data privileges are in the designated areas on the left side of each band. Voice privileges are on the right side of each band. Designated areas between the code and voice privileges have "no privileges" for the General Class operator. These are reserved for grandfathered Advanced Class and current Extra Class operators.

As you can see, grandfathered Advanced and Extra Class operators have the same band privileges that you will enjoy as a General Class operator, they just have a little bit more elbow room. But don't worry – there is plenty of room throughout the General Class voice spectrum for working the world!

160 METERS, 1.8 MHZ - 2.0 MHZ

Your General Class privileges are the same as Advanced and Extra on this band. You may operate voice and code from one end to the other. The 160-meter band is great for long-distance, nighttime communications. It's located just above the AM broadcast radio frequencies. At night, 160 meters lets you work the world.

75/80 METERS, 3.500 MHZ - 4.000 MHZ

On 75/80 meters, General Class code, data, radio teleprinter (RTTY) privileges are from 3.525 to 3.600 MHz, and single sideband voice privileges were recently expanded and go from 3.800 to 4.000 MHz.

60 METERS – 5 CHANNELS

Our 60-meter band was allocated by the FCC on July 3, 2003. Unlike the other ham bands where we have a frequency spectrum, our 60-meter privileges are assigned *5 specific channels* for upper sideband communications. Effective radiated power is limited to 50 watts, and the 5 USB channels buzz with activity. The dialed USB frequencies assigned to the amateur service are: 5330.5 kHz; 5346.5 kHz; 5366.5 kHz; 5371.5 kHz; and 5403.5 kHz. We anticipate increased privileges on the 60-meter band soon.

40 METERS, 7.000 MHZ - 7.300 MHZ

Good day & night

	7.000 MHz	N/T (200 WATTS)		7.175 MHz	7.300 MHz	VOICE SIDEBAND
40 METERS	NO	CODE	NO PRIVILEGES	VOICE		LSB
	7.025 MHz		7.125 MHz			

On 40 meters, General Class code, data, and RTTY privileges are from 7.025 to 7.125 MHz, and single sideband voice privileges recently were expanded and now go from 7.175 to 7.300 MHz. During daylight hours, 40 meters is a great band for base station and mobile contacts up to 500 miles away. At night, 40 meters skips all over the country, and many times all over the world. The 40 meter worldwide foreign broadcast AM shortwave stations that used to cause interference at night have been moved out of this band, opening up the entire band for exclusive amateur radio operations day and night!

30 METERS, 10.100 MHZ - 10.150 MHZ

Hot band for CW & data

	10.100 MHz		10.150 MHz	CW & DATA ONLY
30 METERS		CODE + DATA (200 WATTS)		—

Only code, data, and RTTY are permitted on this band. Thirty meters is located just above the 10-MHz WWV time broadcasts on shortwave radio. Voice is not allowed on this band by any class of amateur operator, and power is limited to 200 watts PEP.

20 METERS, 14.000 MHZ - 14.350 MHZ

Long-range day & evening contacts

	14.000 MHz		14.150 MHz	14.225 MHz	14.350 MHz	VOICE SIDEBAND
20 METERS	NO	CODE	NO PRIVILEGES	VOICE		USB
	14.025 MHz					

This is the best DX worldwide band there is, day or night! Morse code, data, and radioteleprinter (RTTY) privileges extend from 14.025 to 14.150 MHz. General Class voice privileges extend from 14.225 to 14.350 MHz. This is where the real DX activity takes place. Almost 24 hours a day, you should be able to work stations in excess of 5000 miles away with a modest antenna setup on the 20-meter band. If you are a mariner, most of the long range maritime mobile bands are within your privileges as a General Class operator. If you are into recreational vehicles (RVs), there are nets all over the country especially for you. The band "where it's at" is 20 meters when you want to work the world from your car, boat, RV, or home shack!

17 METERS, 18.068 MHZ - 18.168 MHZ

Best during the day

	18.068 MHz	18.110 MHz	18.168 MHz	VOICE SIDEBAND
17 METERS	CODE		VOICE	USB

There is plenty of elbow room here with lots of foreign DX coming in day and night. Most new base antennas have 17 meters included. All emission types are authorized on this newer Amateur Radio band.

15 METERS, 21.000 MHZ - 21.450 MHZ

15 meter General Class CW, data and RTTY privileges extend from 21.025 to 21.200 MHz. Your single sideband voice privileges recently were expanded on 15 meters and go from 21.275 to 21.450 MHz. 15 meters is a great band for daytime skywave contacts, and 15 meters is a popular band for mobile operators because antenna requirements are relatively small. I have worked all over the world on the 15 meter band, mobile.

12 METERS, 24.890 MHZ - 24.990 MHZ

Code, data, and RTTY privileges on the 12-meter band extend from 24.890 to 24.930 MHz. Voice privileges are from 24.930 to 24.990 MHz. You have the same privileges and elbow room as the Advanced and Extra Class operator, too. Although this is a very narrow band, expect excellent daytime range throughout the world. At night, range is limited to groundwave coverage because the ionosphere is not receiving sunlight to produce skywave coverage on this band.

10 METERS, 28.000 MHZ - 29.700 MHZ

General Class CW, data, and RTTY privileges begin at the very bottom of the band, 28.000 MHz, and extend up to 28.300 MHz. Your voice privileges begin immediately at 28.300 MHz and extend up to 29.700 MHz. You have the same privileges on 10 meters as Advanced and Extra Class operators. Recent rule-making now allows all Technician Class operators voice privileges on 10 meters, but only from 28.300 to 28.500 MHz. This will be an exciting area to tune in when the band regularly opens for skywave contacts. This makes the 10 meter band a great spot to talk via skywaves with a new Technician Class operator, with plenty of room above 28.500 for exclusive General, Advanced and Extra Class privileges. And wait until you try the full fidelity of frequency modulation (FM) by giving a quick CQ call on 29.600 MHz, FM! There are even FM repeaters at the top of 10 meters, too.

Don't forget your handheld!
When you upgrade to
general, you keep your VHF/
UHF/SHF privileges.

6 METERS AND UP

Your General Class license allows you unlimited band privileges and unlimited emission privileges on several higher-frequency bands, as shown in *Table 1-1*.

Table 1-1. 6 Meter and Higher Band Privileges

Frequency	Meters
50-54 MHz	6 meters
144-148 MHz	2 meters
222-225 MHz	1.25 meters
420-450 MHz	0.70 meters (70 cm)
902-928 MHz	0.35 meters (35 cm)
1240-1300 MHz	0.23 meters (23 cm)

MICROWAVE BANDS

Your General Class license allows you unlimited band privileges and unlimited emission privileges in the microwave bands, as indicated in *Table 1-2*. These VHF, UHF, and SHF frequencies are the same ones for which you received privileges when you passed your Technician or Technician-Plus Class examinations.

Table 1-2. Microwave Band Frequency Privileges

Frequency	Frequency
2300-2310 MHz	47.0-47.2 GHz
2390-2450 MHz	75.5-1.0 GHz
3.3-3.5 GHz	119.98-20.02 GHz
5.65-5.925 GHz	142-49 GHz
10.0-10.5 GHz	241-50 GHz
24.0-24.25 GHz	All above 300 GHz

My first book, *Technician Class*, for Element 2, provides a detailed explanation of the VHF, UHF, and SHF bands and presents the specific ARRL-recommended band plans for these frequencies. *Technician Class* is available from your local amateur radio dealer, at hamfests, or by calling The W5YI Group at 800/669-9594, or visit **www.w5yi.org**.

THE CONSIDERATE OPERATOR'S FREQUENCY GUIDE

Nothing in the FCC rules recognizes special privileges on any specific frequency for a net, group, or individual. No one "owns" a frequency. Rather, amateur operators rely on "gentlemen's agreements," good amateur practice, and common sense for all ham operators to check to see if the frequency is in use prior to transmitting – regardless of mode.

That said, here's a handy listing of frequencies that are generally recognized for certain modes or activities. My thanks to our friends at *QST* magazine, who gave us permission to reprint it from their March 2007 edition.

Frequency (MHz)	Mode / Activity
1.800 – 2.000	CW
1.800 – 1.810	Digital
1.810	QRP CW calling frequency
1.843-2.000	SSB, SSTV and other wideband modes
1.910	SSB QRP
1.995 – 2.000	Experimental
1.999 – 2.000	Beacons
3.500 – 3.510	CW DX window
3.560	QRP CW calling frequency
3.570 – 3.600	RTTY/Data
3.585 – 3.600	Automatically controlled data stations
3.590	RTTY/Data DX
3.790 – 3.800	DX Window
3.845	SSTV
3.885	AM calling frequency
3.985	QRP SSB calling frequency
7.030	QRP CW calling frequency
7.040	RTTY/Data DX
7.080 – 7.125	RTTY/Data
7.100 – 7.105	Automatically controlled data stations
7.070	PSK
7.285	QRP SSB calling frequency
7.290	AM calling frequency
10.1000 - 10.1300	CW
10.1239	Winlink – Pactor
10.1320	Digital SSTV
10.1345	Olvia, MFSK
10.1386	WSPR
10.1390	JT65 WSJT
10.1405	BPSK31 Propnet
10.1400 - 10.1420	BPSK31
10.1420 - 10.1450	Many digital modes
10.1420 - 10.1440	ALE-400 Hz
10.1470 - 10.1480	PK mail
10.1491 - 10.1495	APRS

Frequency (MHz)	Mode / Activity
14.060	QRP CW calling frequency
14.070 – 14.095	RTTY/Data
14.095 – 14.0995	Automatically controlled data stations
14.100	IBP/NCDXF beacons
14.1005 – 14.112	Automatically controlled data stations
14.230	SSTV
14.285	QRP SSB calling frequency
14.286	AM calling frequency
18.100 – 18.105	RTTY/Data
18.105 – 18.110	Automatically controlled data stations
18.110	IBP/NCDXF beacons
21.060	QRP CW calling frequency
21.070 – 21.110	RTTY/Data
21.090 – 21.100	Automatically controlled data stations
21.150	IBP/NCDXF beacons
21.340	SSTV
21.385	QRP SSB calling frequency
24.920 – 24.925	RTTY/Data
24.925 – 24.930	Automatically controlled data stations
24.930	IBP/NCDXF beacons
28.060	QRP CW calling frequency
28.070 – 28.120	RTTY/Data
28.120 – 28.189	Automatically controlled data stations
28.190 – 28.225	Beacons
28.200	IBP/NCDXF beacons
28.385	QRP SSB calling frequency
28.680	SSTV
29.000 – 29.200	AM
29.300 – 29.510	Satelite downlinks
29.520 – 29.580	Repeater inputs
29.600	FM simplex
29.620 – 29.680	Repeater outputs

TECHNICIAN EXAM ADMINISTRATION

There is one more very important privilege you gain when you achieve General Class status. As a General Class licensee, you may take part in the administration of Element 2 Technician Class written examinations, once you become accredited as a volunteer examiner at the General Class level by a VEC.

So, if you wish to see the amateur service grow in your local area, find 2 other General Class operators, then contact your local or national VEC for accreditation, and start your own testing team for newcomers to our hobby.

> To become an accredited Volunteer Examiner as a General Class licensee call the W5YI VEC at 800-669-9594.

SUMMARY

In late 2006, the FCC "refarmed" code, data, and voice privileges for General, Advanced, and Extra Class operators. All of the band charts in this new book have been updated to reflect the added privileges for high frequency voice operation. The FCC tightened up the code spectrum, and added more elbow room to the voice spectrum. General Class operators gained:

- 50 kHz of voice spectrum on 75 meters
- 50 kHz of voice spectrum on 40 meters
- 25 kHz of voice spectrum on 15 meters

With the elimination of the Morse code test on February 23, 2007, and the addition of all of this voice spectrum, General Class is the place to be!

Are you ready to prepare for the General Class Element 3 written examination? I sure hope so. Welcome – in advance – to the worldwide bands! I hope to hear you on the high frequency bands very soon.

A Little Ham History!

Ham radio has changed a lot in the 100-plus years since radio's inception. In the past 25 years, we have seen some monumental changes! So, before we get started preparing for the exam, I'm going to give you a little refresher lesson about our hobby, its history, and an overview of how you'll progress through the amateur ranks to the top amateur ticket – the Extra Class license. We know this background knowledge will make you a better ham! I'll keep it light and fun, so breeze through these pages.

In this chapter you'll learn all of the licensing requirements under the FCC rules that became effective April 15, 2000. And you'll learn about the six classes of license that were in effect *prior* to those rules changes. That way, when you run into a Novice, Technician Plus or Advanced Class operator on the air, you'll have some understanding of their skill level, experience, and frequency privileges.

WHAT IS THE AMATEUR SERVICE?

There are nearly 700,000 Americans who are licensed amateur radio operators in the U.S. today. The Federal Communications Commission, the Federal agency responsible for licensing amateur operators, defines our radio service this way:

"The amateur service is for qualified persons of all ages who are interested in radio technique solely with a personal aim and without pecuniary interest."

Ham radio is first and foremost a fun hobby! In addition, it is a service. And note the word "qualified" in the FCC's definition – that's the reason why you're studying for an exam; so you can pass the exam, prove you are qualified, and get on the air.

Millions of operators around the world exchange ham radio greetings and messages by voice, teleprinting, telegraphy, facsimile, and television worldwide. Japan, alone, has more than a million hams! It is very commonplace for U.S. amateurs to communicate with Russian amateurs, while China is just getting started with its amateur service. Being a ham operator is a great way to promote international good will.

The benefits of ham radio are countless! Ham operators are probably known best for their contributions during times of disaster. In recent years, many recreational sailors in the Caribbean who have been attacked by modern-day pirates have had their lives saved by hams directing rescue efforts. Following the 9/11 terrorist attacks on the World Trade Center and the Pentagon, literally thousands of local hams assisted with emergency communications. After Katrina blew through New Orleans, ham radio operators were the first to report the levee breach and to warn that flood waters were rushing into the city.

The ham community knows no geographic, political or social barrier. If you study hard and make the effort, you are going to earn your upgrade to General Class and get on the HF worldwide bands. Follow the suggestions in my book and your chances of passing the written exam are excellent, and learning will be easy and fun!

A BRIEF HISTORY OF AMATEUR RADIO LICENSING

Before government licensing of radio stations and amateur operators was instituted in 1912, hams could operate on any wavelength they chose and could even select their own call letters. The Radio Act of 1912 mandated the first Federal licensing of all radio stations and assigned amateurs to the short wavelengths of less than 200 meters. These "new" requirements didn't deter them, however, and within a few years there were thousands of licensed ham operators in the United States.

Since electromagnetic signals do not respect national boundaries, radio is international in scope. National governments enact and enforce radio laws within a framework of international agreements that are overseen by the International Telecommunications Union. The ITU is a worldwide United Nations agency headquartered in Geneva, Switzerland. The ITU divides the radio spectrum into a number of frequency bands, with each band reserved for a particular use. Amateur radio is fortunate to have many bands allocated to it all across the radio spectrum.

In the U.S., the Federal Communications Commission is the government agency responsible for the regulation of wire and radio communications. The FCC further allocates frequency bands to the various services in accordance with the ITU plan – including the Amateur Service – and regulates stations and operators.

By international agreement, in 1927 the alphabet was apportioned among various nations for basic call sign use. The prefix letters K, N and W were assigned to the United States, which also shares the letter A with some other countries.

In the early years of amateur radio licensing in the U.S., the classes of licenses were designated by the letters "A," "B," and "C." The highest license class with the most privileges was "A." In 1951, the FCC dropped the letter designations and gave the license classes names. They also added a new Novice Class – a one-year, non-renewable license for beginners that required a 5-wpm Morse code speed proficiency test and a 20-question written examination on elementary theory and regulations, with both tests taken before one licensed ham.

In 1967, the Advanced Class was added to the Novice, Technician, General and Extra classes. The General exam required 13-wpm code speed, and Extra required 20-wpm. Each of the five written exams were progressively more comprehensive and formed what came to be known as the *Incentive Licensing System.*

In the '70s, the Technician Class license became very popular because of the number of repeater stations appearing on the air that extended the range of VHF and UHF mobile and handheld stations. It also was very fashionable to be able to patch your mobile radio into the telephone system, which allowed hams to make telephone calls from their automobiles long before the advent of cell phones.

In 1979, the international Amateur Service regulations were changed to permit all countries to waive the manual Morse code proficiency requirement for "...stations making use exclusively of frequencies above 30 MHz." This set the stage for the creation of the Technician "no-code" license, which occurred in 1991, when the 5-wpm Morse code requirement for the Technician Class was eliminated. New licensees were now permitted to operate on all amateur bands above 30 MHz. Applicants for the no-code Technician license had to pass the 35-question Novice and 30-question Technician Class written examinations but, for the first time, not a Morse code test. Technician Class amateurs who also passed a 5-wpm code test were awarded a Technician-Plus license. Besides their 30 MHz and higher no-code frequency privileges, Tech-Plus licensees gained the Novice CW privileges and a sliver of the 10 meter voice spectrum.

By this time, there was a total of six Amateur Service license classes – Novice, Technician, Technician-Plus, General, Advanced, and Extra – along with five written exams and three Morse code tests used to qualify hams for their various licenses.

The Amateur Service Is Restructured

As you can see, through the years the ham radio service underwent a myriad of changes. Rules were amended to meet new technology, and the FCC regularly received petitions for "small" rule making. Then, in 1998, the FCC didn't just come up with a new plan for ham radio – they begin working with the ham radio community to restructure our amateur radio service to make it more effective. The objective of this restructuring was to streamline the license process, eliminate unnecessary and duplicated rules, and reduce the emphasis on the Morse code test. The result of this review was a complete restructuring of the US amateur service, which became effective April 15, 2000. The Morse code test for General Class and Extra Class was dropped from 13- and 20-wpm respectively, down to-5 wpm for all. Since 2000, there have been only 3 amateur radio license classes:

• Technician Class, the VHF/UHF entry level license

• General Class, the HF entry level license, which required a 5-wpm code test

• Amateur Extra Class, a technically oriented senior license, based on-5wpm code credit

Individuals with licenses issued before April 15, 2000, have been grandfathered under the new rules. This means that Novice, Technician Plus, and Advanced Class amateurs are able to modify and renew their licenses indefinitely. Technician Plus amateur licenses are renewed as "Technician Class," but their original 5 wpm credit allowed them HF privileges indefinitely.

The Granddaddy of all restructuring occurred on February 23, 2007, when the FCC completely eliminated any Morse code examination for all amateur radio licenses! The FCC action was based on 6,200 written comments, with most supporting the elimination of all code tests.

The FCC also "upgraded" No-Code Technician Class operators for Technician high frequency privileges, mostly code only, on 80 meters, 40 meters, 15 meters, and code, data and 200 kHz of voice spectrum on 10 meters.

In its public notice announcing elimination of the Morse code test requirement, the FCC Commissioners wrote: "We believe that because the international requirement for telegraphy proficiency has been eliminated, we should treat Morse code telegraphy no differently than other amateur service communication techniques. This change eliminates an unnecessary regulatory burden that may discourage current amateur radio operators from advancing their skills and participating more fully in the benefits of amateur radio."

Eliminating the Morse code test in no way diminishes the enthusiasm many ham operators have for the technique of sending dits and dahs over the air. In fact, I believe we will have more General Class hams learning the code than ever before, now that they can do it on high frequency where practicing code with other hams is fun, even though they begin by using only the more simple code characters along with their call signs!

Self-Testing In The Amateur Service

Prior to 1984, all amateur radio exams were administered by FCC personnel at FCC Field Offices around the country. In 1984, the FCC adopted a two-tier system beneath it called the VEC System to handle amateur radio license exams. It also increased the length of the term of amateur radio licenses from five to ten years.

The VEC (Volunteer Examiner Coordinator) System was formed after Congress passed laws that allowed the FCC to accept the services of Volunteer Examiners (or VEs) to prepare and administer amateur service license examinations. The testing activity of VEs is managed by Volunteer Examiner Coordinators (or VECs). A VEC acts as the administrative liaison between the VEs, who administer the various ham examinations, and the FCC, which grants the license.

A team of three VEs, who must be approved by a VEC, is required to conduct amateur radio examinations. General Class amateurs may serve as examiners for the Technician class. Advanced Class amateurs may administer exams for Elements 2 and 3. The Extra Class written Element 4 may only be administered by a VE who holds an Extra Class license.

In 1986, the FCC turned over responsibility for maintenance of the exam questions to the National Conference of VECs, which appoints a Question Pool Committee (QPC) to develop and revise the various question pools according to a schedule. The QPC is required by the FCC to have at least ten times as many questions in each of the pools as may appear on an examination. As a rule, each of the 3 question pools is changed once every 4 years.

The previously-mandated ten written exam topic areas were eliminated and the Question Pool Committee now decides the content of each of the three written examinations. Both the Technician Class Element 2 and General Class Element 3 written examinations contain 35 multiple-choice questions. The Extra Class Element 4 written examination has 50 questions.

OPERATOR LICENSE REQUIREMENTS

To qualify for an amateur operator/primary station license, a person must pass an examination according to FCC guidelines. The degree of skill and knowledge that the candidate demonstrates to the examiners determines the class of operator license for which the person is qualified.

Anyone is eligible to become a U.S. licensed amateur operator (including foreign nationals, if they are not a representative of a foreign government). There is no age limitation – if you can pass the examinations, you can become a ham!

OPERATOR LICENSE CLASSES AND EXAM REQUIREMENTS

Today, there are three amateur operator licenses issued by the FCC – Technician, General, and Extra. Each license requires progressively higher levels of learning and proficiency, and each gives you additional operating privileges. This is known as *incentive licensing* – a method of strengthening the amateur service by offering more radio spectrum privileges in exchange for more operating and electronic knowledge.

There is no waiting time required to upgrade from one amateur license class to another, nor any required waiting time to retake a failed exam. You can even take all three examinations at one sitting if you're really brave! *Table 2-1* details the amateur service license structure and required examinations.

Table 2-1: Current Amateur License Classes and Exam Requirements

License Class	Exam Element	Type of Examination
Technician Class	2	35-question, multiple-choice written examination. Minimum passing score is 26 questions answered correctly (74%).
General Class	3	35-question, multiple-choice written examination. Minimum passing score is 26 questions answered correctly (74%).
Extra Class	4	50-question, multiple-choice written examination. Minimum passing score is 37 questions answered correctly (74%).

ABOUT THE WRITTEN EXAMS

What is the focus of each of the written examinations, and how does it relate to gaining expanding amateur radio privileges as you move up the ladder toward your Extra Class license? *Table 2-2* summarizes the subjects covered in each written examination element.

Table 2-2. Question Element Subjects

Exam Element	License Class	Subjects
Element 2	Technician	Elementary operating procedures, radio regulations, and a smattering of beginning electronics. Emphasis is on VHF and UHF operating.
Element 3	General	HF (high-frequency) operating privileges, amateur practices, radio regulations, and a little more electronics. Emphasis is on HF bands.
Element 4	Extra	Basically a technical examination. Covers specialized operating procedures, more radio regulations, formulas and heavy math. Also covers the specifics on amateur testing procedures.

No Jumping Allowed

All written examinations for an amateur radio license are additive. You *cannot* skip over a license class or by-pass a required examination as you upgrade from Technician to General to Extra. For example, to obtain a General Class license, you must first take and pass the Element 2 written examination for the Technician Class license, plus the Element 3 written examination. To obtain the Extra Class license, you must first pass the Element 2 (Technician) and Element 3 (General) written examinations, and then successfully pass the Element 4 (Extra) written examination.

TAKING THE ELEMENT 3 EXAM

Here's a summary of what you can expect when you go to the session to take the Element 3 written examination for your General Class license. Detailed information about how to find an exam session, what to expect at the session, what to bring to the session, and more, is included in Chapter 4.

Examination Administration

All amateur radio examinations are administered by a team of at least three Volunteer Examiners (VEs) who have been accredited by a Volunteer Examiner Coordinator (VEC). The VEs are licensed hams who volunteer their time to help our hobby grow.

How to Find an Exam Session

Examination sessions are organized under the auspices of an approved VEC. A list of VECs is located in the Appendix on page 203. The W5YI-VEC and the ARRL-VEC are the 2 largest examination groups in the country, and they test in all 50 states. Their 3-member, accredited examination teams are just about *everywhere*. So when you call the VEC, you can be assured they probably have an examination team only a few miles from where you are reading this book right now!

> ***Want to find a test site fast?***
> Visit the W5YI-VEC website at **www.w5yi.org**, or call 800-669-9594.

Taking the Exam

The Element 3 written exam is a multiple-choice format. The VEs will give you a test paper that contains the 35 questions and multiple choice answers, and an answer sheet for you to complete. Take your time! Make sure you read each question carefully and select the correct answer. Once you're finished, double check your work before handing in your test papers.

The VEs will score your test immediately, and you'll know before you leave the exam site whether you've passed. Chances are very good that, if you've studied hard, you'll get that passing grade!

GETTING/KEEPING YOUR CALL SIGN

Once the VE team scores your test and you've passed, the process of getting your official FCC Amateur Radio License begins – usually that same day.

At the exam site, you will complete NCVEC Form 605, which is your application to the FCC for your license. If you pass the exam, the VE team will send on the required paperwork to their VEC. The VEC reviews the paperwork submitted by your exam team and then files your application with the FCC. This filing is done electronically, and your license upgrade will be granted and posted on the FCC's website within a few days.

If you didn't check the "Change Call Sign" box on your NCVEC Form 605 application, you will simply keep your current call sign. However, if you do check the "Change Call Sign" box, you will receive an entry-level "Group D" call sign as if you were a newly-licensed amateur. I suggest that you don't check the "Change Call Sign" box and stick with your present call sign.

As soon as you have your CSCE, you are permitted to go on the air with your General Class HF privileges – even before your paper license arrives in the mail! See Chapter 4 for more details on this process.

Vanity Call Signs

Your first call sign is assigned by the FCC's computer, and you have no choice of letters. However, once you have that call sign, you can apply for a Vanity Call Sign. Again, see Chapter 4 for details.

HOW MANY CLASSES OF LICENSES?

Once you've passed your Element 3 exam and go on the air as a new General Class operator, you'll be talking to fellow hams throughout the U.S. and around the world. Here's a summary of the new and "grandfathered" licenses that your fellow amateurs may hold, and a recap of the level of expertise they have demonstrated in order to gain their licenses.

New License Classes

Following the FCC's restructuring of Amateur Radio that took effect April 15, 2000, there are just three written exams and three license classes – Technician, General, and Extra. But persons who hold licenses issued prior to April 15, 2000, may continue to hold onto their license class and continue to renew it every 10 years for as long as they wish.

"Grandfathered" Licensees

As mentioned previously, individuals licensed prior to April 15, 2000, continue to enjoy band privileges based on their licenses. So, once you get on the air with your new General Class privileges, every now and then you might meet a Novice operator while yakking on 10 meters, or sending CW on 15, 40, or 80 meters.

And then there are the Advanced Class operators who may continue to hold onto their license class designation until they finally decide to move up to Extra Class.

When you look at the brand new Frequency Charts in this book, you'll see that we have updated them to reflect the expanded privileges on many high frequency bands for General Class operators and higher. We also list the sub-band privileges for Extra Class, Advanced Class, General Class, Technician Class and even the Novice. Look for my poster-sized, color-coded chart offered by many ham radio manufacturers that shows all of these operating privileges.

Now here's an interesting twist to the FCC Rules: Part 97.505 of the FCC Amateur Service Rules & Regulations provides credit for General Class Element 3, without examination, to those applicants who can prove they held an expired or unexpired Technician Class license granted before March 21, 1987. Sorry, no other class of license receives this element credit.

"Grandfathered General"

Anyone who received their Technician Class license on or before March 21, 1987, but whose license is now expired, may rejoin the ranks of Amateur Radio by passing only the Element 2 Technician Class examination. When you do that, you also qualify for General Class license without having to take the Element 3 General Class exam!

If you are currently licensed as a Technician and received your original license prior to March 21, 1987, you can submit a copy of your old Technician license and receive credit for the Element 3 exam to qualify for your General Class upgrade. To do so, you must attend a VE testing session, pay the exam fee and submit your documentation and application through the normal VE to VEC session process to receive what is commonly referred to as a General Class "paper upgrade" or "grandfathered General." To see if you qualify, or for more information, contact the W5YI VEC at 800-669-9594.

Gordo and Chip work 500 miles on 432 MHz using just 30 watts!

IT'S EASY!

Probably the primary pre-requisite for passing any amateur radio operator license exam is the will to do it. If you follow my suggestions in this book, your chances of passing the General exam are excellent. And once you pass, then it's on to our *Extra Class* book.

Yes, indeed, 2006 and 2007 FCC rulemaking has opened up all of the high frequency bands for expanded voice privileges, and General Class licensing exams without a Morse code test.

When you become a General Class operator, GET ON THE AIR! Sure, pick up my *Extra Class* book, but start operating as an important step to becoming a good ham and ultimately a candidate for the highest license – Amateur Extra Class. But, GET ON THE AIR as a General Class ham!

Getting Ready for the Exam

Your General Class written examination will consist of 35 multiple-choice questions taken from the 456 questions that make up the 2011-15 Element 3 question pool. Each question on your examination and the multiple-choice answer will be identical to those contained in this book.

This chapter contains the official, complete 456-question FCC Element 3 General Class question pool from which your examination will be taken. Again, your exam will contain 35 of these questions, and you must get 74% of the questions correct – which means you must answer 26 questions correctly in order to pass. This chapter also contains important information about how the exam is constructed using the questions taken from the Element 3 pool.

Your examination will be administered by a team of 3 or more Volunteer Examiners (VEs) – amateur radio operators who are accredited by a Volunteer Examiner Coordinator (VEC). You will receive a Certificate of Successful Completion of Examination (CSCE) when you pass the examination. This is official proof that you have passed the exam and it will be given to you before you leave the exam center.

THE 2011-15 QUESTION POOL

The Element 3 General Class question pool contained in this book is valid from July 1, 2011, through June 30, 2015.

The 456 questions and distracters in the new General Class question pool were developed by the National Conference of Volunteer Examiner Coordinators' Question Pool Committee (NCVEC-QPC). The QPC Chairman is Roland Anders, K3RA, and the QPC members are Perry Green, WY1O, Michael Matson, N6OPH, Larry Pollock, NB5X, and Jim Wiley, KL7CC.

The QPC encourages amateur radio operators throughout the country to submit revised questions for the amateur radio pools to the committee. If you have any suggestions for new or revised questions, you can send them to me, and I'll be happy to forward them on to the QPC. My address is on page 193.

WHAT THE EXAMINATION CONTAINS

The examination questions and the multiple-choice answers (one correct answer and three "distracters") for all license class levels are public information. They are widely published and are identical to those in this book. *There are no "secret" questions.* FCC rules prohibit any examiner or examination team from making any changes to any questions, including any numerical values. No numbers, words, letters, or punctuation marks can be altered from the published question pool. By studying all 456 Element 3 questions in this book, you will be reading the same

ıt will appear on your 35-question Element 3 written examination.
ıf the 456 total questions?

3-1. FCC Element 3 General Class Question Pool

ıc		Total Questions	Exam Questions
G1	Commission's Rules	58	5
G2	Operating Procedures	58	5
G3	Radio Wave Propagation	41	3
G4	Amateur Radio Practices	65	5
G5	Electrical Principles	42	3
G6	Circuit Components	46	3
G7	Practical Circuits	38	3
G8	Signals and Emissions	24	2
G9	Antennas and Feedlines	56	4
G0	Electrical and RF Safety	28	2
TOTALS		456	35

Table 3-1 shows how the Element 3, 35 question examination will be constructed.
The question pool is divided into 10 sub-elements. Each sub-element covers a
different subject, and is divided into topic areas. For example, for the Element 3
examination, two questions from the 35 total exam questions will be taken from sub-
element G 8, Signals and Emissions. On sub-element G1, Commission's Rules, you
will have 5 exam questions on your test. Rusty on electrical principles? Your test
will have only 3 questions from sub-element G 5.

All Volunteer Examination teams use the same multiple-choice question pool.
This uniformity in study material ensures common examinations throughout the
country. Most exams are computer-generated, and the computer selects one question
from each topic area within each subelement for your upcoming Element 3 exam.

Trust me – trust me, every question on your upcoming Element 3 exam will look
very familiar to you by the time you finish studying this book

QUESTION CODING

Each and every question in the 456 question Element 3, General Class pool is
numbered using a **code**. *The coded numbers and letters reveal important facts
about each question!*

The numbering code always contains 5 alphanumeric characters to identify each
question. Here's how to read the question number so you know exactly how the
examination computer will select one question out of each group for your exam.
Once you know this information, you can increase your odds of achieving a "max"
score on the exam, especially if there is a specific group of questions which seems
impossible for you to memorize or understand. When you get to Element 4, Extra
Class – a very tough exam – this trick will really come in handy!

Let's pick a typical question out of the pool – G1A04 – and let me show you how
this numbering code works:

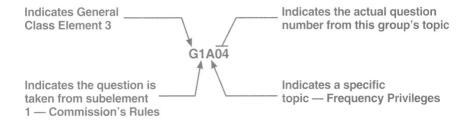

Figure 3-1. Examination Question Coding

- The first character "G" identifies the license class question pool from which the question is taken. "T" would be for Technician. "G" is for General, and "E" would be for Extra.

- The second digit, a "1", identifies the subelement number, 1 through 0. General subelement 1 deals with FCC Rules.

- The third character, "A", indicates the topic area within the subelement. Topic "A" deals with your General Class frequency privileges.

- The fourth and fifth digits indicate the actual question number within the subelement topic's group. The "04" indicates this is the fourth question about frequency privileges, and within the topic area of Commission's Rules. There are 15 individual questions in topic area G1A, *but only one question out of this topic group will appear on the test.*

Here's the Secret Study Hint

Only one exam question will be taken from any single group! A computer-generated test is set up to take only one question from each topic group. It cannot skip any one group, nor can it take any more than one question from that group.

Your upcoming Element 3, General Class written exam is relatively easy with no bone-crusher math formulas. But when you get to the *Extra Class* book, there may be one or two groups that have formulas so complicated that you may want to wait until the very end to digest them. And if you decide to skip them completely, guess what – how many questions out of any one group? That's right, only one per group. This means you are not going to get hammered on any upcoming test with a whole bunch of questions dealing with a specific topic. Great secret, huh?

Study Time

How long will it take you to prepare for your upcoming exam?

The General Class question pool in this book is valid from July 1, 2011, through June 30, 2015. It contains a total of 456 questions – but don't panic! Most questions are repeated several different ways, and these "repeats" reinforce what you've already learned. It is probably going to take about 30 days to work through this book and prepare for your General Class exam. Remember, you must take the written exams in order: Element 2 Technician, Element 3 General, and Element 4 Extra.

QUESTIONS REARRANGED FOR SMARTER LEARNING

The first thing you'll notice when you look at how the Element 3 question pool is presented in this book is that I have *completely rearranged the entire General Class question pool* to precisely follow my weekend ham radio training classes. This rearrangement will take you *logically* through each and every one of the 456 Q & A's. The questions are arranged here by 18 topic areas in a way that eliminates the need for you to jump back and forth between topic groups or subelements to match up questions on similar topics.

For example, I have taken all of the questions in the pool that talk about where you can operate your ham radio and grouped them together into one area that allows you to better understand all of the material that relates to this topic. This arrangement of the Q&A's follows a natural learning process beginning with your new General Class privileges, participating on a VE team, and getting on the air, and ending with questions on antennas, feed lines, and RF safety.

Trust me, the reorganization of all of the test questions in the 456 pool has been tested and finely-tuned in hundreds of my weekend classes. This method of learning WORKS! You will probably cut the amount of study time in half simply by following the question pool from front to back as presented here in my book!

Let me assure that each and every Element 3 Q&A is in this book. A cross-reference of all 456 questions is found on pages 210 to 211 in the back of the book, along with the syllabus used by the Question Pool Committee to develop this new Element 3 pool.

This book – and my General Class audio course – contain all 456 General Class questions, 4 possible answers, the noted correct answer, and my upbeat description of how the correct answer works into the real world of amateur radio. We highlight **KEY WORDS** that will help you remember the correct answer and provide you with a fast review of the entire question pool just before you sit for the big General exam.

We also include many web addresses that can provide you with hours of fascinating study on selected "hot topics" that will help you really understand the real world of ham radio behind the Q & A's. You can visit the site while you study, or visit them after you've earned your General Class license and are on the air.

When you visit some of these websites, it may not be immediately apparent why we are suggesting that you go there. Some addresses take you to the sites of local ham clubs. Most of these are specialty clubs, and they contain lots of information on how to operate on repeaters, or satellites, or provide educational resources on learning about electronics or antennas or how radios work.

And a disclaimer – while we have worked hard to make sure all of these addresses are current at the time of publication, websites move, addresses change, or sites simply go away. So if you find an address that doesn't work, feel free to drop us an e-mail so we can update this for the next printing of our book. And if you know of a website that you think is a gem, then send us that information and we'll consider it for our next printing.

Here's our e-mail address:
masterpubl@aol.com

How to Read the Questions

Using an actual question from page 27, here is a guide to explain what it is you will be studying as you go through all of the Q&As in the book:

Official
Q&A

G1C05 Which of the following is a limitation on transmitter power on the 28 MHz band?
A. 100 watts PEP output.   C. 1500 watts PEP output.
B. 1000 watts PEP output.   D. 2000 watts PEP output.
Although Novice and Technician Class operators are restricted to 200 watts on high frequency bands, as a General Class operator you could run up to *1500 watts* PEP (Peak Envelope Power) output. [97.313] **ANSWER C.**

FCC Part   Correct   Key
97 Rule   Answer   Words To
Citation   Remember

Topic Areas

Here is a list of my 18 topic areas showing the page where it starts in the book. Again, there is a complete cross reference list of the Q&As in numerical order on pages 210 to 211 in the Appendix, along with the official Question Pool Syllabus.

STUDY SUGGESTIONS

Finally, as you get ready to start studying the questions, here are some suggestions to make your learning easier:

1. Read over each multiple-choice answer carefully. Some answers start out looking good, but turn bad during the last 2 or 3 words. If you speed read the answers, you could very easily go for a wrong answer because you didn't read them all the way through. Also, don't count on the multiple-choice answers always appearing in the exact same A-B-C-D order on your actual computer-generated test. While they won't change any words in the answers, they will sometimes scramble the A-B-C-D order.

2. Keep in mind that there is only one question on your test that will come from each group, and track how many groups in each sub-element.

3. Give this book to a friend, and ask him or her to read you the correct answer. You then reply with the question wording.

4. Mark the heck out of your book! When the pages begin to fall out, you're probably ready for the exam!

5. My book is available on audio CDs, too. So if you'd like to listen to me read the questions and answers to you while you're driving your car, riding your bike, or laying on the beach, I can do that for you. The CD symbol, disk number, and track number at the beginning of each topic section keys this book to the audio book. You can get my book on audio CD where you purchased this book, or by calling 800-669-9594, or by visiting www.w5yi.org.

CD 1 TRACK 2

6. Highlight the keywords one week before the test. Then speed read the brightly highlighted key words twice a day before the exam.

Are you ready to work through the 456 Q & A's? Put a check mark by the easy ones that you may already know the answer for, and put a little circle by any question that needs a little bit more study. Save your highlighting work until a few days before your upcoming test. Work the Q & A's for about 30 minutes at a time. I'll drop in a little bit of humor to keep you on track; and if you actually need my live words of encouragement, you can call me on the phone Monday through Thursday, 10:00 a.m. to 4:00 p.m. (California time), 714-549-5000.

THE QUESTION POOL, PLEASE

Okay, this is the big moment – your General Class, Element 3, question pool. Don't freak out and get overwhelmed with the prospect of learning 456 Q & A's. You will find that the topic content is repeated many times, so you're really going to breeze through this test without any problem!

Your Passing CSCE

G1D09 How long is a Certificate of Successful Completion of Examination (CSCE) valid for exam element credit?

A. 30 days.
B. 180 days.
C. 365 days.
D. For as long as your current license is valid.

See how confident we are that you will soon pass your General upgrade exam? When you pass you exam, your examiners will all sign a Certificate of Successful Completion of Examination (CSCE). You are good to go on the air immediately with your new privileges! This CSCE is proof-positive that you passed the exam, just in case there is a paperwork snafu down the line. Your license usually is upgraded within two weeks, so the *365-day* period of this *CSCE* applies only in RARE cases where paperwork gets lost. [97.9(b)] **ANSWER C.**

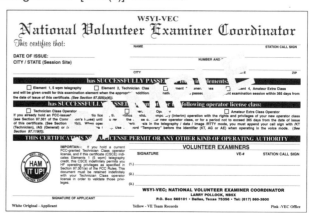

Don't lose your CSCE – it is proof of your privileges until your license arrives.

G1D06 When must you add the special identifier "AG" after your call sign if you are a Technician Class licensee and have a CSCE for General Class operator privileges, but the FCC has not yet posted your upgrade on its Web site?

    A. Whenever you operate using General Class frequency privileges.
    B. Whenever you operate on any amateur frequency.
    C. Whenever you operate using Technician frequency privileges.
    D. A special identifier is not required as long as your General Class license application has been filed with the FCC.

Until your call sign appears on the Universal Licensing System data base, *add the special identifier "slash AG"* when you operate using your new General Class privileges. You don't need to append this identifier when operating with your current Technician frequency privileges, just when using your new General Class frequencies. [97.119(f)(2)] **ANSWER A.**

G1D01 Which of the following is a proper way to identify when transmitting using phone on General Class frequencies if you have a CSCE for the required elements but your upgrade from Technician has not appeared in the FCC database?

    A. Give your call sign followed by the words "General Class."
    B. No special identification is needed.
    C. Give your call sign followed by "slant AG."
    D. Give your call sign followed the abbreviation "CSCE."

When you upgrade from Technician Class to General Class it will take about 10-15 days for your upgrade to appear on the Universal Licensing System database. But you don't need to wait to go on the air, because you have your CSCE (Certificate of Successful Completion of Examination) that allows you to *give your current* Technician Class *call sign followed by the words "slant AG."* Everyone will welcome you as a new General Class operator on the air when they hear "slant AG." [97.119(f)(2)] **ANSWER C.**

G1D03 On which of the following band segments may you operate if you are a Technician Class operator and have a CSCE for General Class privileges?

    A. Only the Technician band segments until your upgrade is posted on the FCC database.
    B. Only on the Technician band segments until your license arrives in the mail.
    C. On any General or Technician Class band segment.
    D. On any General or Technician Class band segment except 30 and 60 meters.

Once you pass your General Class written examination, you are good to *go on the air on all* of the *General Class* segments of the worldwide *bands* immediately with the CSCE, *plus your present Technician frequencies*. Keep my book handy to spot the operating segments for General Class operation on each band. [97.9(b)] **ANSWER C.**

Just earned your General Class CSCE?
Go ahead and get on the air!

G1E09 What language must you use when identifying your station if you are using a language other than English in making a contact using phone emission?

A. The language being used for the contact.
B. Any language if the US has a third party agreement with that country.
C. English.
D. Any language of a country that is a member of the ITU.

Great band conditions these days – tomorrow morning, early, expect Europe to come blasting in on the 20 and 15 meter bands. You hook-up with a chap in Italy, and you are fluent in Italian. Go for it! It's perfectly acceptable on skywave communications to speak the language of the amateur operator you have met on the air. Just be sure to *identify with your call sign in English* every 10 minutes, and when you sign off. Do it in English! [97.119(b)(2)] **ANSWER C.**

When you pass your exam, you'll receive
your CSCE and lots of congratulations!

Your New General Bands

 **Elmer Point:** *Here are the formulas that you need to convert frequency to wavelength, and wavelength to frequency:*

Converting Frequency to Wavelength

To find wavelength (λ) in meters, if you know frequency (f) in megahertz (MHz), Solve:

$$\lambda(meters) = \frac{300}{f(MHz)}$$

Converting Wavelength to Frequency

To find frequency (f) in megahertz (MHz), if you know wavelength (λ) in meters, Solve:

$$f(MHz) = \frac{300}{\lambda(meters)}$$

G1A01 On which of the following bands is a General Class license holder granted all amateur frequency privileges?

A. 60, 20, 17, and 12 meters.

B. 160, 80, 40, and 10 meters.

C. 160, 60, 30, 17, 12, and 10 meters.

D. 160, 30, 17, 15, 12, and 10 meters.

As a new General, you will gain high frequency privileges on each and every ham band. But until you upgrade to Extra, you don't necessarily get ALL of that band's frequency privileges. Extra and Advanced operators have some exclusive room on 80, 40, 20, and 15 meters. Not to worry – *General Class has all frequency privileges on 160, 30, 17, 12, and 10 meters*, and on the five 60 meter channels. [97.301(d), 97.303(s)] **ANSWER C.**

G1A11 Which of the following frequencies is available to a control operator holding a General Class license?
A. 28.020 MHz.
B. 28.350 MHz.
C. 28.550 MHz.
D. All of these choices are correct.

As a General Class operator, you receive all privileges from 28.000 MHz to 29.700 MHz on the 10 meter band. *All of the choices are correct.* [97.301(d)] **ANSWER D.**

	28.500 MHz				VOICE SIDEBAND
10 METERS	N/T (200 WATTS)			29.700 MHz	USB
	CODE		VOICE	FM	
	28.000 MHz 28.300 MHz			29.500 MHz	

G1C05 Which of the following is a limitation on transmitter power on the 28 MHz band?
A. 100 watts PEP output.
B. 1000 watts PEP output.
C. 1500 watts PEP output.
D. 2000 watts PEP output.

Although Novice and Technician Class operators are restricted to 200 watts on high frequency bands, as a General Class operator you could run up to *1500 watts* PEP (Peak Envelope Power) output. [97.313] **ANSWER C.**

G1E02 When may a 10 meter repeater retransmit the 2 meter signal from a station having a Technician Class control operator?
A. Under no circumstances.
B. Only if the station on 10 meters is operating under a Special Temporary Authorization allowing such retransmission.
C. Only during an FCC declared general state of communications emergency.
D. Only if the 10 meter repeater control operator holds at least a General Class license.

10 meter repeaters may only operate between 29.5 to 29.7 MHz. Technician Class operators have no privileges on these frequencies. However, as a General, you can set up your home station as a cross-band relay system. This could allow a Technician on the 2-meter band to end up transmitting and receiving on the 10-meter band. Think of all the excitement you can give Technician Class operators on 2 meters when the 10 meter FM segment of the band is open for worldwide skywave communications. This is perfectly legal to do, providing you stay at the control point of your station. You may never leave the relay station unattended without a *General Class, or higher, control operator* on active duty at the control point. [97.205(a)] **ANSWER D.**

G1A06 Which of the following frequencies is in the 12 meter band?
A. 3.940 MHz.
B. 12.940 MHz.
C. 17.940 MHz.
D. 24.940 MHz.

Your new General Class *12 meter band* extends from 24.890 to 24.990 MHz, and there is only one answer with 24 MHz. Remember, to convert meters to megahertz, or megahertz to meters, 300 divided by either megahertz or meters will get you close! 300 divided by 12 = 25, which is close enough for the *24.940 MHz* answer. [97.301(d)] **ANSWER D.**

	24.890 MHz	24.930 MHz	24.990 MHz	VOICE SIDEBAND
12 METERS	CODE	VOICE		USB

G1C02 What is the maximum transmitting power an amateur station may use on the 12 meter band?
    A. 1500 PEP output, except for 200 watts PEP output in the Novice portion.
    B. 200 watts PEP output.
    C. 1500 watts PEP output.
    D. An effective radiated power equivalent to 50 watts from a half-wave dipole.
On the *12 meter band*, you can run the legal limit of *1500 watts* PEP output.
[97.313(a)(b)] **ANSWER C.**

G1A10 Which of the following frequencies is within the General Class portion of the 15 meter band?
    A. 14250 kHz.   C. 21300 kHz.
    B. 18155 kHz.   D. 24900 kHz.
The *15 meter band* is at 21 MHz (21000 kHz), so *21300 kHz* is great place for you to operate voice to regularly work the world in the morning and afternoon.
[97.301(d)] **ANSWER C.**

15 METERS	21.025 MHz	N/T (200 WATTS)		21.275 MHz	21.450 MHz	VOICE SIDEBAND
	NO	CODE	NO PRIVILEGES	VOICE		USB
	21.000 MHz		21.200 MHz			

G1A08 Which of the following frequencies is within the General Class portion of the 20 meter phone band?
    A. 14005 kHz.   C. 14305 kHz.
    B. 14105 kHz.   D. 14405 kHz.
Remember that phone privileges for high frequency General Class operation are those at the top of the band. The *20 meter band* ends at *14350 kHz*, so 14305 kHz is within General Class voice privileges. [97.301(d)] **ANSWER C.**

20 METERS	14.000 MHz		14.150 MHz	14.225 MHz	14.350 MHz	VOICE SIDEBAND
	NO	CODE	NO PRIVILEGES	VOICE		USB
	14.025 MHz					

G1C04 Which of the following is a limitation of power on the 14 MHz band?
    A. Only the minimum power necessary to carry out the desired communications should be used.
    B. Power must be limited to 200 watts when transmitting between 14.100 MHz and 14.150 MHz.
    C. Power should be limited as necessary to avoid interference to another radio service on the frequency.
    D. Effective radiated power cannot exceed 3000 watts.
Although you might be permitted to run an amplifier with 1500 watts output, always try to run the *minimum power necessary to make contact* with the other station. This will keep you out of your neighbors TV and HiFi systems. [97.313]
ANSWER A.

G1A02 On which of the following bands is phone operation prohibited?
    A. 160 meters.   C. 17 meters.
    B. 30 meters.   D. 12 meters.

As a new General, you can operate throughout the entire 30 meter band using CW and DATA emissions, 300 baud, with power output limited to 200 watts. *No phone (voice) allowed on 30 meters*. [97.305] **ANSWER B.**

	10.100 MHz	10.150 MHz	
30 METERS		CODE + DATA (200 WATTS)	**CW & DATA ONLY**

G1A03 On which of the following bands is image transmission prohibited?

A. 160 meters.
B. 30 meters.
C. 20 meters.
D. 12 meters.

Image transmissions include popular slow scan television and facsimile. On *30 meters, image transmissions are prohibited*. [97.305] **ANSWER B.**

G1C01 What is the maximum transmitting power an amateur station may use on 10.140 MHz?

A. 200 watts PEP output.
B. 1000 watts PEP output.
C. 1500 watts PEP output.
D. 2000 watts PEP output.

On the *30 meter* CW and Data *band*, we can only run *200 watts* peak envelope power output to conform to FCC rules. [97.313(c)(1)] **ANSWER A.**

G1A05 Which of the following frequencies is in the General Class portion of the 40 meter band?

A. 7.250 MHz.
B. 7.500 MHz.
C. 40.200 MHz.
D. 40.500 MHz.

Welcome to *40 meters*, where we recently gained an additional 50 kHz of voice spectrum, from 7175 to 7300 kHz, lower sideband. Look for me most weekday mornings on *7.250 MHz*, within the General Class portion of the 40 meter band. [97.301(d)] **ANSWER A.**

	7.000 MHz	N/T (200 WATTS)		7.175 MHz	7.300 MHz	**VOICE SIDEBAND**
40 METERS	NO	CODE	NO PRIVILEGES	VOICE		**LSB**
	7.025 MHz		7.125 MHz			

G1E03 In what ITU region is operation in the 7.175 to 7.300 MHz band permitted for a control operator holding an FCC-issued General Class license?

A. Region 1.
B. Region 2.
C. Region 3.
D. All three regions.

As a General Class operator, you will likely be operating on our continent in ITU Region 2. This allows voice privileges, as well as data and CW, from 7.175 to 7.300 MHz (7175 to 7300 kHz), but only in ITU Region 2. If you head off to the South Seas in ITU Region 3, your 40 meter privileges will be different. Here in the US, including Hawaii, we are in *ITU Region 2*. [97.301(d)] **ANSWER B.**

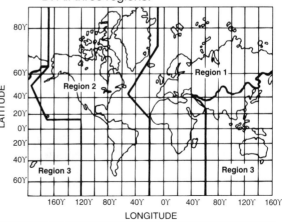

ITU Regions

G1A04 **Which of the following amateur bands is restricted to communication on only specific channels, rather than frequency ranges?**
A. 11 meters.
B. 12 meters.
C. 30 meters.
D. 60 meters.

We gained a new ham band, 60 meters, shared with a few government stations on a non-interference basis – they can interfere with us, but we cannot interfere with them! We are authorized *5 discreet*, upper sideband voice *channels*, and these are great frequency assignments for passing regional radio traffic. *60 meters* is the only band where we have been assigned specific channel allocations, and we may only use upper sideband voice on each of these channels. No Morse code, and no data, either. [97.303(s)] **ANSWER D.**

60 METERS	5 SPECIFIC CHANNELS – 50 WATTS ERP					VOICE SIDEBAND
	5330.5 kHz	5346.5 kHz	5366.5 kHz	5371.5 kHz	5403.5 kHz	USB

G1C03 **What is the maximum bandwidth permitted by FCC rules for Amateur Radio stations when transmitting on USB frequencies in the 60 meter band?**
A. 2.8 kHz.
B. 5.6 kHz.
C. 1.8 kHz.
D. 3 kHz .

On the 60 meter band, there are five channels allocated on specific frequencies, with a maximum bandwidth of 2.8 kHz. This is 2.8 kHz UPPER sideband, no lower sideband allowed on 60 meters. On *60 meters*, only UPPER sideband is allowed, and it's also a good idea to turn off your speech processor so you don't exceed our *2.8 kHz bandwidth limitation*. [97.303(s)] **ANSWER A.**

G2D07 **Which of the following is required by the FCC rules when operating in the 60 meter band?**
A. If you are using other than a dipole antenna, you must keep a record of the gain of your antenna.
B. You must keep a log of the date, time, frequency, power level and stations worked.
C. You must keep a log of all third party traffic.
D. You must keep a log of the manufacturer of your equipment and the antenna used.

FCC rules require 60 meter band, 5-channel operation not to exceed 50 watts effective radiated power out as measured on a dipole antenna. On a dipole, gain is zero, so 50 watts into the dipole from your transmitter will NOT exceed 50 watts effective radiated power output. However, *if you're transmitting using a home-made beam*, you will need to turn your power output down to correspond with the amount of gain the beam may exhibit. If the beam offers 3 dB gain in the forward direction, this 2 times increase in effective radiated power will require you to reduce your transmitter power output to -3 dB, or half of 50 watts. You then must *note this in a station logbook* and keep it as a permanent record as part of your station written files. [97.303(s)] **ANSWER A.**

G1A15 **What is the appropriate action if, when operating on either the 30 or 60 meter bands, a station in the primary service interferes with your contact?**
A. Notify the FCC's regional Engineer in Charge of the interference.
B. Increase your transmitter's power to overcome the interference.
C. Attempt to contact the station and request that it stop the interference.
D. Move to a clear frequency.

Both our 30 meter and 60 meter bands are shared, and authorized government users always have priority. So *change frequency* or stop transmitting. [97.303] **ANSWER D.**

G1A14 Which of the following applies when the FCC rules designate the Amateur Service as a secondary user on a band?
    A. Amateur stations must record the call sign of the primary service station before operating on a frequency assigned to that station.
    B. Amateur stations are allowed to use the band only during emergencies.
    C. Amateur stations are allowed to use the band only if they do not cause harmful interference to primary users.
    D. Amateur stations may only operate during specific hours of the day, while primary users are permitted 24 hour use of the band.

Our relatively new 60 meter channelized ham band is shared with a few government agencies. They are primary on the band, which means *we must not cause harmful interference to government users*. [97.303] **ANSWER C.**

G1A07 Which of the following frequencies is within the General Class portion of the 75 meter phone band?
    A. 1875 kHz.   C. 3900 kHz.
    B. 3750 kHz.   D. 4005 kHz.

The 75/80 meter ham band is a good one for short-range, daytime communications and long-range, nighttime skywave signals. 80 meters refers to the bottom of the band for CW and data, and *75 meters* refers to the top of the band for phone. 300 divided by 75 = 4.000 MHz. To convert MHz to kHz, move the decimal point 3 places to the right. *3900 kHz* is within the phone privileges for General Class. [97.301(d)] **ANSWER C.**

G1A09 Which of the following frequencies is within the General Class portion of the 80 meter band?
    A. 1855 kHz.   C. 3560 kHz.
    B. 2560 kHz.   D. 3650 kHz.

General Class CW and data privileges for the *80 meter band* extend from 3525 kHz through *3600 kHz*. 3560 kHz is a good spot for CW, RTTY, and DATA. [97.301(d)] **ANSWER C.**

G1C06 which of the following is a limitation on transmitting power on the 1.8 MHz band?
    A. 200 watts PEP output.   C. 1200 watts PEP output.
    B. 1000 watts PEP output.   D. 1500 watts PEP output.

As a new General Class operator, you are permitted to run up to *1500 watts* of Peak Envelope Power on all bands other than 60 meters and 30 meters. However, always run the minimum power necessary. [97.313] **ANSWER D.**

G1A12 When General Class licensees are not permitted to use the entire voice portion of a particular band, which portion of the voice segment is generally available to them?
  A. The lower frequency end.
  B. The upper frequency end.
  C. The lower frequency end on frequencies below 7.3 MHz and the upper end on frequencies above 14.150 MHz.
  D. The upper frequency end on frequencies below 7.3 MHz and the lower end on frequencies above 14.150 MHz.

Study our band plan charts and see where your privileges begin and end. *Voice privileges*, called "phone," are always at the *top of the bands*. (Okay, 60 meters has its specific 5 channels of voice from top to bottom!) [97.301] **ANSWER B.**

Voice (phone) privileges are always at the top end of the bands.

G2B08 What is the "DX window" in a voluntary band plan?
  A. A portion of the band that should not be used for contacts between stations within the 48 contiguous United States.
  B. An FCC rule that prohibits contacts between stations within the United States and possessions on that band segment.
  C. An FCC rule that allows only digital contacts in that portion of the band.
  D. A portion of the band that has been voluntarily set aside for digital contacts only.

It is important to begin your General Class worldwide band operation in accordance with voluntary band plans. Most high frequency bands have a spot where stateside hams will call and listen for *ONLY foreign country DX stations*. On the 160 meter band, 1830 to 1850 kHz is long recognized as the DX WINDOW – no idle chit-chatting here! Use the high frequency "DX WINDOW" as a great spot to listen for worldwide DX. **ANSWER A.** ☞ Visit: www.arrl.org/bandplans

G1A13 Which, if any, amateur band is shared with the Citizens Radio Service?
  A. 10 meters.   C. 15 meters.
  B. 12 meters.   D. None.

If you know a ham operator who has been licensed more than 50 years, he will tell you war stories about how we lost our 11 meter ham band to the Citizens Radio Service. *We now share NO bands with CB*. [97.303] **ANSWER D.**

Elmer Point: *As a brand new General, spend a few days on the air LISTENING before you go on the air. Look at this band plan, and then tune in to hear how everything has its place on the radio dial. One of the best ways to complete your first transmission using your new General Class privileges is to answer an upbeat CQ call. Double-check to make sure you are within your General Class privileges, and check the band plan to make sure you are not answering a station looking only for foreign DX.*

160 Meters (1.8 – 2.0 MHz)

1.800 - 2.000	CW
1.800 - 1.810	Digital Modes
1.810	CW QRP
1.843 - 2.000	SSB, SSTV and other wideband modes
1.910	SSB QRP
1.995 - 2.000	Experimental
1.999 - 2.000	Beacons

80 Meters (3.5 – 4.0 MHz)

3.590	RTTY / Data DX
3.570 - 3.600	RTTY / Data
3,790 - 3.800	DX window
3.845	SSTV
3.885	AM calling frequency

40 Meters (7.0 – 7.3 MHz)

7.040	RTTY / Data DX
7.080 - 7.125	RTTY / Data
7.171	SSTV
7.290	AM calling frequency

30 Meters (10.1 – 10.15 MHz)

10.100 - 10.1300	CW
10.130 - 10.140	RTTY
10.140 - 10.150	Packet
10.1239	Winlink – Pactor
10.1320	Digital SSTV
10.1345	Olvia, MFSK
10.1386	WSPR
10.1390	JT65 WSJT
10.1405	BPSK31 Propnet
10.1400 - 10.1420	BPSK31
10.1420 - 10.1450	Many digital modes
10.1420 - 10.1440	ALE-400 Hz
10.1470 - 10.1480	PK mail
10.1491 - 10.1495	APRS

20 Meters (14.0 – 14.35 MHz)

14.070 - 14.095	RTTY
14.095 - 14.0995	Packet
14.100	NCDXF Beacons
14.1005 - 14.112	Packet
14.230	SSTV
14.286	AM calling frequency

17 Meters (18.068 – 18.168 MHz)

18.100 - 18.105	RTTY
18.105 - 18.110	Packet

15 Meters (21.0 – 21.45 MHz)

21.070 - 21.110	RTTY / Data
21.340	SSTV

12 Meters (24.89 – 24.99 MHz)

24.920 - 24.925	RTTY
24.925 - 24.930	Packet

10 Meters (28.0 – 29.7 MHz)

28.000 - 28.070	CW
28.070 - 28.150	RTTY
28.150 - 28.190	CW
28.200 - 28.300	Beacons
28.300 - 29.300	Phone
28.680	SSTV
29.000 - 29.200	AM
29.300 - 29.510	Satellite Downliknks
29.520 - 29.590	Repeater Inputs
29.600	FM Simplex
29.610 - 29.700	Repeater Outputs

G2D08 Why do many amateurs keep a log even though the FCC doesn't require it?

　　A. The ITU requires a log of all international contacts.

　　B. The ITU requires a log of all international third party traffic.

　　C. The log provides evidence of operation needed to renew a license without retest.

　　D. To help with a reply if the FCC requests information.

Whenever I let a guest ham talk over my equipment, I usually write down details of who the other operator was in my *station log*. This way, *if the FCC should ask* who was the control operator or third party during a transmission on a given date and time, I might be able to look it up and have a clue about what went out over the airwaves! **ANSWER D.**

☞ **Visit:**
www.w3beinformed.org

While the FCC doesn't require it, keeping a log of your station operation is a good idea.

G2D09 What information is traditionally contained in a station log?

　　A. Date and time of contact.

　　B. Band and/or frequency of the contact.

　　C. Call sign of station contacted and the signal report given.

　　D. All of these choices are correct.

I *write down all of my station operation in the log book*, and this way my notes are continuously available in case I need to go back and figure out why I got this QSL card from someone who said they worked me out on the boat. I usually write down my latitude and longitude while operating mobile marine, and this way I can send them a card back with all the details found in my log. **ANSWER D.**

G2D01 What is the Amateur Auxiliary to the FCC?
    A. Amateur volunteers who are formally enlisted to monitor the airwaves for rules violations.
    B. Amateur volunteers who conduct amateur licensing examinations.
    C. Amateur volunteers who conduct frequency coordination for amateur VHF repeaters.
    D. Amateur volunteers who use their station equipment to help civil defense organizations in times of emergency.

The *Amateur Auxiliary* is made up of *volunteer hams* who are formally enlisted to *monitor the airwaves for rule violations*, and who report violations to the local FCC Compliance and Information Bureau office. **ANSWER A.**

G2D02 Which of the following are objectives of the Amateur Auxiliary?
    A. To conduct efficient and orderly amateur licensing examinations.
    B. To encourage amateur self regulation and compliance with the rules.
    C. To coordinate repeaters for efficient and orderly spectrum usage.
    D. To provide emergency and public safety communications.

The benchmark of the amateur service is *self-regulation and compliance* with the rules. Hams help other hams stay on the straight and narrow. The FCC has little time nor budget to monitor the amateur bands for rules violations. **ANSWER B.**

G2D03 What skills learned during "hidden transmitter hunts" are of help to the Amateur Auxiliary?
    A. Identification of out of band operation.
    B. Direction finding used to locate stations violating FCC Rules.
    C. Identification of different call signs.
    D. Hunters have an opportunity to transmit on non-amateur frequencies.

On a transmitter hunt, portable, mobile, and base station beam antennas are used to triangulate and home-in on a specific signal. It takes 3 or more stations to triangulate, and this develops cooperative skills with other operators. Homing-in on a signal requires techniques in using attenuators, to lessen signal strength when you are right on top of the transmitter. You can foxhunt with almost any simple transceiver, and if you really get hooked you'll increase your skill in *using directional antennas to track down almost any signal on the air*. **ANSWER B.**

W6JAY practices his fox hunt direction-finding skills.

Elmer Point: *Amateur Auxiliary operators know the Part 97 Rules and Regulations inside and out. You should, too! The FCC rules are written in plain language providing fun reading to see all that the FCC encourages you to do on the air. Call 800-669-9594, and tell them Gordo wants you to read the FCC Part 97 Rule book.*

G1B11 How does the FCC require an amateur station to be operated in all respects not specifically covered by the Part 97 rules?
A. In conformance with the rules of the IARU.
B. In conformance with Amateur Radio custom.
C. In conformance with good engineering and good amateur practice.
D. All of these choices are correct.

When hams come up with new radio signaling techniques, they might not be found in the FCC rules. However, if this new signaling technique conforms to *good engineering and good amateur practice* you can begin operating, even though the rules may not specifically authorize this new particular type of radio emission. [97.101(a)] **ANSWER C.**

G1B12 Who or what determines "good engineering and good amateur practice" as applied to the operation of an amateur station in all respects not covered by the Part 97 rules?
A. The FCC.   C. The IEEE.
B. The Control Operator.   D. The ITU.

Only *the Federal Communications Commission (FCC)* can decide what meets "good engineering and good amateur practice." [97.101(a)] **ANSWER A.**

G1B08 When choosing a transmitting frequency, what should you do to comply with good amateur practice?
A. Review FCC Part 97 Rules regarding permitted frequencies and emissions.
B. Follow generally accepted band plans agreed to by the Amateur Radio community.
C. Before transmitting, listen to avoid interfering with ongoing communication.
D. All of these choices are correct.

Before you transmit with your new General Class privileges, ask yourself the following:
Am I within my General Class privileges?
Am I operating in accordance with the band plan?
Is anyone else using the frequency?

Keep the frequency privileges chart in the back of my book handy, along with manufacturer color band plan charts. This way, you will meet the Part 97 Rules, and likely meet thousands of friendly hams. **ANSWER D.**

G1B05 When may music be transmitted by an amateur station?

A. At any time, as long as it produces no spurious emissions.
B. When it is unintentionally transmitted from the background at the transmitter.
C. When it is transmitted on frequencies above 1215 MHz.
D. When it is an incidental part of a manned space craft retransmission.

About the only *music* you will ever hear on the ham bands is limited to some of the audio feeds *from the space shuttle or international space station retransmissions* where they sometimes wake up the crew with reveille, or sing "Happy Birthday" to them in outer space. All other forms of music are not allowed. [97.113(a)(5),(e)] **ANSWER D.**

☞ **Visit: www.amsat.org**
www.work-stat.com

Space Shuttle
Photo courtesy of N.A.S.A.

G1B06 When is an amateur station permitted to transmit secret codes?

A. During a declared communications emergency.
B. To control a space station.
C. Only when the information is of a routine, personal nature.
D. Only with Special Temporary Authorization from the FCC.

We encourage you to join AMSAT – Radio Amateur Satellite Corporation, a nonprofit scientific organization that supports our ham satellite programs. Specific AMSAT ham stations are allowed to transmit secure *secret codes to control* ham radio equipment located on *satellites*. [97.113(a)(4) and 97.207(f)] **ANSWER B.**

G1B07 What are the restrictions on the use of abbreviations or procedural signals in the Amateur Service?

A. Only "Q" codes are permitted.
B. They may be used if they do not obscure the meaning of a message.
C. They are not permitted.
D. Only "10 codes" are permitted.

Common abbreviations on the ham bands *do not obscure the meaning of our comms,* and such phrases as "73," "QRZ?," or "Please QSL," are perfectly acceptable. Popular Q signals are given on page 56. The following is a list of common prowords. [97.113(a)(4)] **ANSWER B.**

Proword	Meaning	Proword	Meaning
Affirmative	Yes	Number	Message number (in numerals) follows
All after	Say again all after _____		
All before	Say again all before _____	Out	End of transmission, no answer required or expected
Break	Used to separate message heading, text and ending	Over	End of transmission, answer is expected. Go ahead. Transmit.
Break	Stop transmitting		
Correct	That is correct	Roger	I have received your transmission satisfactorily
Figures	Numerals follow		
From	Originator follows	Say again	Repeat
Groups	Numeral(s) indicating number of text words follows	Slant	Slant bar
		This is	This transmission is from the station whose call sign follows
Incorrect	That is incorrect		
Initial	Single letter follows	Time	File time or date-time group of the message follows
I say again	I repeat		
I spell	Phonetic spelling follows	To	Addressee follows
Message follows	A message which requires recording follows	Wait	Short pause
		Wait out	Long pause
More to follow	I have more traffic for you	Word after	Say again word after _____
Negative	No, not received	Word before	Say again word before _____

MARS – Army Radiotelephone Prowords

G1B09 When may an amateur station transmit communications in which the licensee or control operator has a pecuniary (monetary) interest?

A. When other amateurs are being notified of the sale of apparatus normally used in an amateur station and such activity is not done on a regular basis.
B. Only when there is no other means of communications readily available.
C. When other amateurs are being notified of the sale of any item with a monetary value less than $200 and such activity is not done on a regular basis.
D. Never.

Ham radio "swap nets" are a great way to look for, or to *sell, used amateur radio equipment* over the ham airwaves. All equipment must be ham radio related – no selling your Granddad's Model T. [97.113(a)(3)] **ANSWER A.**

ELMER HINT *Recently, the Federal Communications Commission clarified its rules regarding ham stations at your place of employment. Rules do not permit us to use ham radio in place of commercial two way radios for daily business activities, but ham radio disaster preparation nets are an exception. If the purpose of the short weekly net is to support the community in a time of disaster, and as long as the ham radio system is not directly supporting your company's business venture, your company's ham radio emergency operation center may be good to go on the air for weekly community disaster preparedness training.*

School teachers, on the clock, are also permitted to use ham radio in the classroom, even though they are getting paid for their instruction.

Here's the bottom line – don't use ham radio to further your business, or add to your company's bottom line. There are plenty of other radio services for commercial comms, and even non-licensed radio services such as Multi Use Radio Service, Part 15 spread spectrum radios, and Family Radio Service are a perfect way to support your operation outside of the ham bands. Ham radio is NOT a replacement for a company radio system.

G2D05 When is it permissible to communicate with amateur stations in countries outside the areas administered by the Federal Communications Commission?

A. Only when the foreign country has a formal third party agreement filed with the FCC.

B. When the contact is with amateurs in any country except those whose administrations have notified the ITU that they object to such communications.

C. When the contact is with amateurs in any country as long as the communication is conducted in English.

D. Only when the foreign country is a member of the International Amateur Radio Union.

As a new General Class operator, you can work hundreds of countries and become an expert in DX, plus learn a little bit of a foreign language! *As long as NEITHER THEIR nor OUR administrations have notified the International Telecommunications Union that ham communications are forbidden*, you are good to go work any foreign ham in any foreign country. [97.111 (a) (1)] **ANSWER B.**

G1E01 Which of the following would disqualify a third party from participating in stating a message over an amateur station?

A. The third party's amateur license had ever been revoked.

B. The third party is not a U.S. citizen.

C. The third party is a licensed amateur.

D. The third party is speaking in a language other than English, French, or Spanish.

You are not allowed to let a "third party" talk over your equipment if they were previously licensed as a ham and their *license has been revoked*. Don't let them "con" you into letting them speak over the microphone! [97.115(b)(2)] **ANSWER A.**

G1E05 What types of messages for a third party in another country may be transmitted by an amateur station?

A. Any message, as long as the amateur operator is not paid.

B. Only messages for other licensed amateurs.

C. Only messages relating to Amateur Radio or remarks of a personal character, or messages relating to emergencies or disaster relief.

D. Any messages, as long as the text of the message is recorded in the station log.

During a third party traffic exchange, make sure that the station in the other country only transmits *personal messages*. In an emergency, they are allowed to transmit messages relating to *disaster relief*. [97.115(a)(2), 97.117] **ANSWER C.**

Elmer Point: Before you allow third party traffic at your station, make sure your guest operator understands the rules: no business; no profanity; no secret codes; no music, and only a language that you can understand. It's also a good idea to keep a logbook with details of the third party conversation.

List of Countries Permitting Third-Party Traffic

Antigua and Barbuda..........V2	El Salvador.......................YS	Paraguay.............................ZP
Argentina......................LU	The Gambia C5	Peru....................................OA
Australia ....VK	Ghana................................9G	Philippines.........................DU
Austria, Vienna...........4U1VIC	Grenada J3	Pitcairn Island VR6
Belize.............................V3	Guatemala........................TG	St. Christopher & Nevis V4
Bolivia............................CP	Guyana.............................8R	St. Lucia J6
Bosnia-Herzegovina T9	Haiti ....HH	St. Vincent & Grenadines .. J8
Brazil ....PY	Honduras ....HR	Sierra Leone......................9L
Canada.................VE, VO, VY	Israel.................................4X	South Africa ....ZS
Chile ....CE	Jamaica............................6Y	Swaziland......................3D6
Colombia........................HK	Jordan ....JY	Trinidad and Tobago 9Y
Comoros D6	Liberia..............................EL	Turkey ....TA
Costa Rica ....TI	Marshall Is.......................V6	United Kingdom...............GB
Cuba...............................CO	Mexico.............................XE	Uruguay CX
Dominica.........................J7	Micronesia.......................V6	Venezuela.........................YV
Dominican Republic ....HI	Nicaragua........................YN	ITU-Geneva..............4U1ITU
Ecuador........................HC	Panama............................HP	VIC-Vienna................4U1VIC

G1E07 With which foreign countries is third party traffic prohibited, except for messages directly involving emergencies or disaster relief communications?

A. Countries in ITU Region 2.

B. Countries in ITU Region 1.

C. Every foreign country, unless there is a third party agreement in effect with that country.

D. Any country which is not a member of the International Amateur Radio Union (IARU).

This question asks where third party traffic is prohibited. Except in an emergency, we are *prohibited from passing international third party traffic unless we have a third party agreement in effect with that country*. [97.115(a)(2)] **ANSWER C.**

G1E08 Which of the following is a requirement for a non-licensed person to communicate with a foreign Amateur Radio station from a station with an FCC-granted license at which a licensed control operator is present?

A. Information must be exchanged in English.

B. The foreign amateur station must be in a country with which the United States has a third party agreement.

C. The control operator must have at least a General Class license.

D. All of these choices are correct.

To pass *third party traffic* to a station in another country, we must have a *third party agreement* with that country. [97.115(a)(b)] **ANSWER B.**

Elmer Point: Want to operate HF while travelling abroad? The following CEPT countries allow U.S. Amateurs to operate in their countries without a reciprocal license. Be sure to carry a copy of your FCC license along with a copy of FCC Public Notice DA99-1098 with you in case someone asks if you have approval to operate in their country

Austria	Greenland	Norway
Belgium	Hungary	Portugal
Bosnia & Herzegovina	Iceland	Romania
Bulgaria	Ireland	Slovak Republic
Croatia	Italy	Slovenia
Cyprus	Latvia	Spain
Czech Republic	Liechtenstein	Sweden
Denmark	Lithuania	Switzerland
Estonia	Luxembourg	Turkey
Faroe Islands	Monaco	United Kingdom & its
Finland	Montenegro	possessions
France & its possessions	Netherlands	
Germany	Netherlands Antilles	

Going to Europe?
U.S. Amateurs with a General Class license will be granted CEPT Novice Radio Amateur License privileges in accordance with ECC Recommendation (05)06 (as amended), which can be found on the internet at www.erodocdb.dk/doks/implement_doc_adm.aspx?docid=2136.

Be a VE!

G1D02 **What license examinations may you administer when you are an accredited VE holding a General Class operator license?**

    A. General and Technician.
    B. General only.
    C. Technician only.
    D. Extra, General and Technician.

As a new General Class operator, I hope you will apply for Volunteer Examiner accreditation. You may need a Contact Volunteer Examiner to sponsor you. As a General Class operator, you and 2 other certified VE General Class (or higher) operators could administer an Element 2 *Technician Class* written examination. [97.509(b)(3)(i)] **ANSWER C.**

G1D05 **Which of the following is sufficient for you to be an administering VE for a Technician Class operator license examination?**

    A. Notification to the FCC that you want to give an examination.
    B. Receipt of a CSCE for General Class.
    C. Possession of a properly obtained telegraphy license.
    D. An FCC General Class or higher license and VEC accreditation.

To take part in an examination session, you will need to hold a minimum of an *FCC General Class amateur license*, and you must be *accredited by a volunteer examiner coordinator (VEC)*. And remember, it takes a minimum of 3 accredited examiners to be present to conduct the test session. We hope you will soon become an accredited examiner because we need more exam givers for the Element 2 Technician written exam. [97.509(b)(3)(i)] **ANSWER D.**

G1D07 **Volunteer Examiners are accredited by what organization?**

    A. The Federal Communications Commission.
    B. The Universal Licensing System.
    C. A Volunteer Examiner Coordinator.
    D. The Wireless Telecommunications Bureau.

Individual Volunteer *Examiners* are *accredited by a Volunteer Examiner Coordinator*. The largest VECs are the American Radio Relay League (ARRL) and the W5YI-VEC. [97.509(b)(1)] **ANSWER C.**

G1D04 Which of the following is a requirement for administering a Technician Class operator examination?

A. At least three VEC accredited General Class or higher VEs must be present.

B. At least two VEC accredited General Class or higher VEs must be present.

C. At least two General Class or higher VEs must be present, but only one need be VEC accredited.

D. At least three VEs of Technician Class or higher must be present.

As a General Class operator, you have the ability to participate in the volunteer examination system. A contact VE may need to sign off on your VE application. That application goes to the Volunteer Exam Coordinator. Once you are approved, *you and 2 other accredited General Class and higher volunteer examiners* may administer Technician Class (Element 2) operator examinations. [97.509(a)(b)] **ANSWER A.**

G1D08 Which of the following criteria must be met for a non-U.S. citizen to be an accredited Volunteer Examiner?

A. The person must be a resident of the U.S. for a minimum of 5 years.

B. The person must hold an FCC-granted Amateur Radio license of General Class or above.

C. The person's home citizenship must be in the ITU 2 region.

D. None of these choices is correct; non-U.S. citizens cannot be volunteer examiners.

Non U.S. citizens may become both ham operators with a U.S.A. license, as well as volunteer examiners if they *hold an FCC-granted General Class license* or higher. [97.509(b)(3)] **ANSWER B.**

G1D10 What is the minimum age that one must be to qualify as an accredited Volunteer Examiner?

A. 12 years.　　　　　C. 21 years.

B. 18 years.　　　　　D. There is no age limit.

Eighteen years of age or older for Volunteer Examiner accreditation. Remember, while there is no age limit to obtain a ham license, *Volunteer Examiners must be at least 18 years old*. [97.509(b)(2)] **ANSWER B.**

Kids make great ham radio operators, but you have to be 18 or older to become a VE.

Elmer Point: *Our listing of Volunteer Exam Coordinators will lead you to a phone number of your local area testing team. Look for it in the Appendix on page 203. Testing teams are always looking for examinees, so they'll be delighted to hear that you want to take an exam. And once you're a General, you may want to join them as an examiner!* ☞ **Visit: www.w5yi-vec.org**

Elmer Point: *"Roger on your QTH, and fine business on your new rig. Hope you can QSL our QSO, and I wish you very seven three." Say what? Huh? Here's a glossary with a sampling of ham radio lingo that you'll hear when you're on the air!*

CQ CQ CQ this is a station on HF looking for anyone to have a nice friendly contact
CQDX CQDX CQDX this is a station calling ONLY for a contact to a foreign station, not looking for a USA contact
Old Man this really doesn't mean you are old, but rather a term for a fellow ham radio operator
YL usually an unmarried lady
XYL usually a married lady
Harmonics your children
73 best regards to you
88 an affectionate hug to a YL or XYL
Ham a licensed amateur radio operator (with every ham having his or her own idea of this term's origin)
ARRL American Radio Relay League, our number one, non-profit organization that you should join.
APRS Automatic Position Reporting System - automatic GPS radio tracking, usually on 10 MHz HF
DX a station a LONG way away
Break only use this word to break into a conversation with emergency or priority traffic.
     To enter a conversation, politely, only by using your callsign, never "Break".
CQ contest you can answer this call if you are familiar with the precise "exchange" the other operator is looking for.
QRZed who is the station calling me?
Down 10 move down in frequency 10 kHz
5 9 9 your signal report is strong, loud, and clear!
In the Mud your signal is extremely weak
QRM interference from another station on a close frequency
QTH your station location
QSL please send me a QSL card   (also...I agree)
QRN power line or automobile static
QRU does anyone have traffic for me?
QRP low power station
QRT going off the air
QRX stand by - I need to do something
QSV your signal is fading in and out
QSO a communications contact
QST calling all ham radio operators
QSY we need to move off this frequency
TVI interfering with a television set
Heil a premier after-market microphone system
HF worldwide high frequency bands
MARS Military Affiliate Radio Service
Nets regular frequency meeting spot at a certain time for all interested hams
OO an official observer monitoring for rule violations
My Shack your home radio location
Mobile in motion driving down the road with a big HF rig
Ragchew having a long winded conversation
RFI every time I transmit, my windshield wipers self-activate!
Handi-ham an active radio ham who has overcome physical challenges
Splatter maybe turn down your microphone gain
Traffic an incoming message for you
HI HI radio laughter on CW

Voice Operation

G2B06 What is a practical way to avoid harmful interference when selecting a frequency to call CQ on CW or phone?

A. Send "QRL?" on CW, followed by your call sign; or, if using phone, ask if the frequency is in use, followed by your call sign.

B. Listen for 2 minutes before calling CQ.

C. Send the letter "V" in Morse code several times and listen for a response.

D. Send "QSY" on CW or if using phone, announce "the frequency is in use," then send your call and listen for a response.

Soon you will be a General Class operator on the high frequency airwaves. Sometimes, propagation only allows you to hear half a conversation – you might not hear another station transmitting only a couple hundred of miles away. So, a good way to double check that a frequency is clear for your intended CQ is to *send "QRL?" on CW*, followed by your call sign, or *on phone simply ask, "Is the frequency in use?"* followed by your call sign. This will make you one great ham! **ANSWER A.**

G2B07 Which of the following complies with good amateur practice when choosing a frequency on which to initiate a call?

A. Check to see if the channel is assigned to another station.

B. Identify your station by transmitting your call sign at least 3 times.

C. Follow the voluntary band plan for the operating mode you intend to use.

D. All of these choices are correct.

Before placing a call on the worldwide bands, listen for about one minute to make sure the specific frequency is open. Next, double *check that you are following the band plan* and are transmitting within your privileges. Then, give your call sign and ask if the frequency is in use. Do this about 3 times. NOW, you are relatively assured that the frequency is open for you to call CQ. **ANSWER C.**

G2B01 Which of the following is true concerning access to frequencies?
   A. Nets always have priority.
   B. QSO's in process always have priority.
   C. No one has priority access to frequencies, common courtesy should be a guide.
   D. Contest operations must always yield to non-contest use of frequencies.

High frequency nets most always take place on the exact published net frequency. But what happens if that frequency is already in use by some brand new General Class hams who don't realize a net takes place on that frequency every day? The courteous net controller would politely interrupt and encourage the new Generals to join in on the net that "is soon to begin on this frequency." But if the other two stations are unable to hear the request or don't wish to move, the only option is to conduct the net on a nearby clear frequency. As a new General, *be a courteous ham*, and always relinquish your frequency for a published and popular net.
ANSWER C.

G2A08 Which of the following is a recommended way to break into a conversation when using phone?
   A. Say "QRZ" several times followed by your call sign.
   B. Say your call sign during a break between transmissions from the other stations.
   C. Say "Break. Break. Break." and wait for a response.
   D. Say "CQ" followed by the call sign of either station.

When you hear a conversation between two hams, a polite way to join this QSO (communication) is to simply *say your call sign in between one station turning it over to the other station*. Say your call sign in a cheerful way, making it sound like you wish to enter the conversation in a friendly way. Don't just blurt out your call sign – sound pleasant, as if asking permission to join in on the QSO. **ANSWER B.**

G2A11 What does the expression "CQ DX" usually indicate?
   A. A general call for any station.
   B. The caller is listening for a station in Germany.
   C. The caller is looking for any station outside their own country.
   D. A distress call.

The term "CQ" is used by hams to "fish" for a new station to answer their call.

Say your "CQ" with a smile and sound excited about making the call to anyone hearing you. A drab, lifeless "CQ" is like fishing with old bait. If you hear someone calling *"CQ DX,"* this means they are not looking for just any stateside contact, but rather they are *seeking calls only from very distant or foreign stations* thousands of miles away. These usually are very experienced operators, so stay tuned and learn from the "pros" how to call out and get rare "DX" responses. **ANSWER C.**

This DX station, on the top of a hill, will enjoy some great contacts.

G4A03 What is normally meant by operating a transceiver in "split" mode?
A. The radio is operating at half power.
B. The transceiver is operating from an external power source.
C. The transceiver is set to different transmit and receive frequencies.
D. The transmitter is emitting a SSB signal, as opposed to DSB operation.

Almost all HF transceivers have a *"split" mode* that allows you to *listen on one frequency* and instantly *transmit on another*. On older equipment, you are working with VFO-A and VFO-B. On newer equipment, some split-mode features may actually let you listen to your own split transmit frequency, while at the same time receiving the incoming signal. Foreign DX operators like to work split because it allows them to create a "window" where they will listen several kHz higher within our own authorized ham band for US stations. An example might be a foreign station transmitting on 14.145, while listening for USA General Class voice calls on 14.227. We listen down on 14.145, but transmit on the frequency that they are listening to, 14.227! Operating split requires some learning time, so be patient and, most important, only transmit in OUR band where the foreign station says they will be listening UP. Do not transmit on the foreign station's frequency as they may be operating outside of our band while listening UP several kHz! **ANSWER C.**

G4A12 Which of the following is a common use for the dual VFO feature on a transceiver?
A. To allow transmitting on two frequencies at once.
B. To permit full duplex operation, that is transmitting and receiving at the same time.
C. To permit ease of monitoring the transmit and receive frequencies when they are not the same.
D. To facilitate computer interface.

The dual VFO feature on the modern transceiver with same-band receive allows you to *listen on one frequency and simultaneously listen on another* when operating split. With the right radio, you can even add a professional headset from Heil Sound that will put one frequency in your left ear and another frequency in your right ear! That's how the pro DXers work split! **ANSWER C.**

G2B03 If propagation changes during your contact and you notice increasing interference from other activity on the same frequency, what should you do?
A. Tell the interfering stations to change frequency.
B. Report the interference to your local Amateur Auxiliary Coordinator.
C. As a common courtesy, move your contact to another frequency.
D. Increase power to overcome interference.

On the worldwide General Class ham bands, propagation will sometimes cause stations on the same frequency to all of a sudden come in right on top of your ongoing contact. Be a good ham and *move your contact to another frequency*, if you can, to avoid the interference. **ANSWER C.**

G2B05 What is the customary minimum frequency separation between SSB signals under normal conditions?
A. Between 150 and 500 Hz.   C. Approximately 6 kHz.
B. Approximately 3 kHz.   D. Approximately 10 kHz.

When operating single sideband, your emission will take up approximately 3 kHz of bandwidth. Always *stay at least 3 kHz away from any other station* on the air that is using an adjacent frequency. **ANSWER B.**

G4D10 How close to the lower edge of the 40 meter General Class phone segment should your displayed carrier frequency be when using 3 kHz wide LSB?
　　A. 3 kHz above the edge of the segment.
　　B. 3 kHz below the edge of the segment.
　　C. Your displayed carrier frequency may be set at the edge of the segment.
　　D. Center your signal on the edge of the segment.
On the 40 meter band, we would want to operate *lower sideband*, and transmit at least *3 kHz above our band edge* of 7.175 MHz. **ANSWER A.**

G4D08 What frequency range is occupied by a 3 kHz LSB signal when the displayed carrier frequency is set to 7.178 MHz?
　　A. 7.178 to 7.181 MHz.　　　　C. 7.175 to 7.178 MHz.
　　B. 7.178 to 7.184 MHz.　　　　D. 7.1765 to 7.1795 MHz.
7.178 MHz is on the 40 meter band where we usually operate lower sideband, which means our signal will extend DOWN 3 kHz from 7.178 MHz. We would then occupy *7.175 to 7.178 MHz*. **ANSWER C.**

G4D11 How close to the upper edge of the 20 meter General Class band should your displayed carrier frequency be when using 3 kHz wide USB?
　　A. 3 kHz above the edge of the band.
　　B. 3 kHz below the edge of the band.
　　C. Your displayed carrier frequency may be set at the edge of the band.
　　D. Center your signal on the edge of the band.
On the 20 meter band, *USB, stay at least 3 kHz below the top edge of the band*. Don't transmit voice any higher than 14.347 MHz. **ANSWER B.**

G4D09 What frequency range is occupied by a 3 kHz USB signal with the displayed carrier frequency set to 14.347 MHz?
　　A. 14.347 to 14.647 MHz.　　　C. 14.344 to 14.347 MHz.
　　B. 14.347 to 14.350 MHz.　　　D. 14.3455 to 14.3485 MHz.
On the 20 meter band (14.347 MHz) our upper sideband signal will extend UP 3 kHz, so we will occupy *14.347 up to 14.350 MHz*. **ANSWER B.**

G4A11 Which of the following is a use for the IF shift control on a receiver?
　　A. To avoid interference from stations very close to the receive frequency.
　　B. To change frequency rapidly.
　　C. To permit listening on a different frequency from that on which you are transmitting.
　　D. To tune in stations that are slightly off frequency without changing your transmit frequency.
You are tuned into my "Gordo net" on 7250 kHz. About 3 kHz away is another conversation, slightly "bleeding over" on your receiver. Not to worry. When the Gordo net begins, simply adjust the IF shift control on your radio, and magically the *repositioning of the IF pass band will help minimize the sounds of the other station* just a few kilohertz away. Some transceivers may also offer an additional pass band control to further home-in on the signals you want to hear. **ANSWER A.**

G2D10 What is QRP operation?

A. Remote piloted model control.
B. Low power transmit operation.
C. Transmission using Quick Response Protocol.
D. Traffic relay procedure net operation.

Many ham operators on worldwide frequencies enjoy operating *QRP* – the Q code for *low power operation*. Hams get a big kick out of working all the way around the world with less power than that used by a tiny night-light 4 watt bulb. You can look for QRP operation at specific spots on the radio dial band plan. **ANSWER B.**

G2A10 Which of the following statements is true of SSB VOX operation?

A. The received signal is more natural sounding.
B. VOX allows "hands free" operation.
C. Frequency spectrum is conserved.
D. Provides more power output.

Most worldwide radios have a voice-operated relay circuit, abbreviated *"VOX."* If you have a big base station microphone, this is a neat *"hands-free"* circuit to minimize you having to reach over to depress the push-to-talk switch. Just be sure you never leave the VOX circuit turned on when you are not right at your turned-on equipment! **ANSWER B.**

☞ **Visit: www.heilsound.com**

Using a headset with an attached mike on VOX will keep both hands free when you're taking on the worldwide bands while at your home station.

G1E04 Which of the following conditions require an Amateur Radio station licensee to take specific steps to avoid harmful interference to other users or facilities?

A. When operating within one mile of an FCC Monitoring Station.
B. When using a band where the Amateur Service is secondary.
C. When a station is transmitting spread spectrum emissions.
D. All of these choices are correct.

A relatively new form of ham radio emission is called "spread spectrum" and you must take specific steps to avoid harmful interference to other stations. Also, when operating on 30 meters and 60 meters on High Frequency, we are secondary on the frequencies and must not cause interference. And, if you live within one mile of an FCC monitoring station, take steps to avoid any harmful interference. You do NOT want the FCC knocking on your door. *All of these choices are correct*.
[97.13(b), 97.311(b), 97.303] **ANSWER D.**

G1E06 Which of the following applies in the event of interference between a coordinated repeater and an uncoordinated repeater?

A. The licensee of the non-coordinated repeater has primary responsibility to resolve the interference.

B. The licensee of the coordinated repeater has primary responsibility to resolve the interference.

C. Both repeater licensees share equal responsibility to resolve the interference.

D. The frequency coordinator bears primary responsibility to resolve the interference.

If you decide to put up your own repeater, better have a big bank account! They are expensive. Also, seek coordination, because a *non-coordinated repeater has primary responsibility* to resolve interference to another repeater on the same frequency. [97.205(c)] **ANSWER A.**

G1E10 What portion of the 10 meter band is available for repeater use?

A. The entire band.

B. The portion between 28.1 MHz and 28.2 MHz.

C. The portion between 28.3 MHz and 28.5 MHz.

D. The portion above 29.5 MHz.

Chances are that your new high frequency transceiver has a selection for FM, frequency modulation. The only place that FM is allowed for high frequency operation is at the very top of the 10 meter band. The portion above 29.5 MHz does permit FM, with FM repeater **output** on the following frequencies:

> 29.620
> 29.640
> 29.660
> 29.680

Each **input** is 100 kHz lower

> 29.520
> 29.540
> 29.560
> 29.580

The 10 meter *FM repeater* Simplex frequency is 29.600, no offset. Switch over to FM to see what you hear at the top of the 10 meter band, *above 29.500 MHz.* [97.205 (b)] **ANSWER D.**

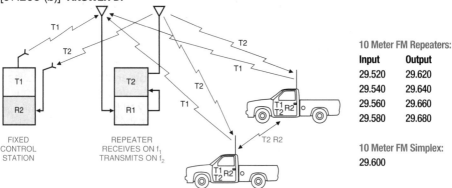

10 Meter FM Repeaters:	
Input	**Output**
29.520	29.620
29.540	29.640
29.560	29.660
29.580	29.680

10 Meter FM Simplex:

29.600

Repeater

Source: *Mobile 2-Way Radio Communications,* G. West, Copyright ©1992 Master Publishing, Inc., Niles, IL

G8B05 Why isn't frequency modulated (FM) phone used below 29.5 MHz?
A. The transmitter efficiency for this mode is low.
B. Harmonics could not be attenuated to practical levels.
C. The wide bandwidth is prohibited by FCC rules.
D. The frequency stability would not be adequate.
We cannot use frequency modulation phone *below 29.5 MHz* because the *bandwidth is simply too wide*. Most FM simplex operation is found on 29.6 MHz, a good spot to enjoy your new General Class privileges. There are also repeaters up in this range. **ANSWER C.**

G2A05 Which mode of voice communication is most commonly used on the high frequency amateur bands?
A. Frequency modulation.   C. Single sideband.
B. Double sideband.   D. Phase modulation.
When you gain your new privileges on High Frequency, the majority of voice modes are *single sideband, SSB*. **ANSWER C.**

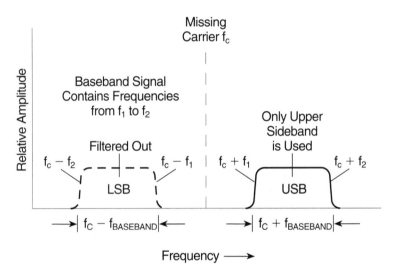

SSB signals are Amplitude Modulated (AM)
with the carrier and one sideband suppressed.

G2A06 Which of the following is an advantage when using single sideband as compared to other analog voice modes on the HF amateur bands?
A. Very high fidelity voice modulation.
B. Less bandwidth used and higher power efficiency.
C. Ease of tuning on receive and immunity to impulse noise.
D. Less subject to static crashes (atmospherics).
The advantages of *single sideband* are that it *occupies less spectrum* than double-sideband AM, *is power efficient*, and there is no continuous carrier in between syllables in your transmission. **ANSWER B.**

G2A07 Which of the following statements is true of the single sideband (SSB) voice mode?
A. Only one sideband and the carrier are transmitted; the other sideband is suppressed.
B. Only one sideband is transmitted; the other sideband and carrier are suppressed.
C. SSB voice transmissions have higher average power than any other mode.
D. SSB is the only mode that is authorized on the 160, 75 and 40 meter amateur bands.

When we transmit using *SSB*, only a single sideband, approximately 2.8 kHz wide, is sent out over the air. The *opposite sideband is suppressed*. There also is *no carrier* in a properly-adjusted SSB signal. This means your radio gets a complete rest during each break in your syllables! This saves power and is great for battery operation in the field. **ANSWER B.**

G2A04 Which mode is most commonly used for voice communications on the 17 and 12 meter bands?
A. Upper sideband.   C. Vestigial sideband.
B. Lower sideband.   D. Double sideband.

Since *17 and 12 meters* are higher in frequency than 20 meters, we always use *upper sideband*. **ANSWER A.**

G2A01 Which sideband is most commonly used for voice communications on frequencies of 14 MHz or higher?
A. Upper sideband.   C. Vestigial sideband.
B. Lower sideband.   D. Double sideband.

14 MHz is the 20 meter band. On 20 meters, and on 17, 15, 12, and 10 meters, plus the 5 channels on the new 60 cm band, plus VHF and UHF weak signal operating, we use *upper sideband*. Believe it or not, except for 60 meters, there is no rule violation for using lower sideband where everyone else normally is using upper sideband. But be a good operator and choose upper sideband on frequencies from 14 MHz and higher, including 60 meters.
ANSWER A.

☞ Visit: www.ac6v.com/nets.htm

When you're using voice — even when you're operating portable on the beach — make sure to use the correct sideband.

G2A03 Which of the following is most commonly used for SSB voice communications in the VHF and UHF bands?
A. Upper sideband.   C. Vestigial sideband.
B. Lower sideband.   D. Double sideband.

When you upgrade to General Class, we hope you will continue to stay active on the *VHF and UHF* bands, too. If you operate weak signal on VHF and UHF, our voice mode is always *upper sideband*. **ANSWER A.**

G2A02 Which of the following modes is most commonly used for voice communications on the 160, 75, and 40 meter bands?

A. Upper sideband.
B. Lower sideband.
C. Vestigial sideband.
D. Double sideband.

We use *lower sideband (LSB) on 160, 75, and 40 meters*. And while it is not absolutely illegal to use lower sideband on 20 meters and shorter wavelength bands, good operating procedure would always indicate that you should "go with the flow" and use the proper sideband. **ANSWER B.**

G2A09 Why do most amateur stations use lower sideband on the 160, 75 and 40 meter bands?

A. Lower sideband is more efficient than upper sideband at these frequencies.
B. Lower sideband is the only sideband legal on these frequency bands.
C. Because it is fully compatible with an AM detector.
D. Current amateur practice is to use lower sideband on these frequency bands.

Remember that the choice of upper and lower sideband is by gentleman's agreement, and you won't find much about it, except for 60 meters USB, in the FCC rule book. On the *160, 75, and 40 meter bands, amateur "practice" is to use lower sideband*. Most modern amateur worldwide high-frequency equipment automatically selects upper sideband for the bands 20 meters and up, and lower sideband for the bands 40 meters and down. **ANSWER D.**

SIDEBAND	FREQUENCY BAND IN METERS							
USB			60		20	17	15	12	10
LSB	160	75/80		40					

Sideband Usage on Amateur Radio Bands

Elmer Point: Want to learn more about how radios work? I suggest you read and study **Basic Communications Electronics** by Jack Hudson, W9MU, and Jerry Luecke, KB5TZY. It explains how transmitters and receivers work, the science behind antennas, integrated circuits, and more. You can pick up a copy at your ham radio store, online at www.w5yi.org, or by calling W5YI Group at 800-669-9594. Understanding how radios work will add to your enjoyment of your General Class privileges!

☞ Visit: www.dxzone.com/catalog/Operating_Modes

CW Lives!

G2B04 When selecting a CW transmitting frequency, what minimum frequency separation should you allow in order to minimize interference to stations on adjacent frequencies?

A. 5 to 50 Hz.   C. 1 to 3 kHz.

B. 150 to 500 Hz.   D. 3 to 6 kHz.

When *operating CW*, try to separate yourself from other CW transmissions by at least *150 to 500 Hz*. This will give your CW signal a distinct tonal difference from the other station, and hopefully will not cause interference. **ANSWER B.**

G2C06 What does the term "zero beat" mean in CW operation?

A. Matching the speed of the transmitting station.

B. Operating split to avoid interference on frequency.

C. Sending without error.

D. Matching your transmit frequency to the frequency of a received signal.

When a net control station asks everyone to *"zero beat"* their CW operation, they would like you to *match their frequency* so that everyone's CW tone will sound about the same. **ANSWER D.** ☞ **Visit: http://aa9pw.com/morsecode**

G2C05 What is the best speed to use answering a CQ in Morse Code?

A. The fastest speed at which you are comfortable copying.

B. The speed at which the CQ was sent.

C. A slow speed until contact is established.

D. 5 wpm, as all operators licensed to operate CW can copy this speed.

Ready to try your first CW CQ? Send at a relatively slow rate, and expect that any other station will send at this slow rate in their response to your CQ. Same thing applies in reverse. If you hear a station sending CQ at a ridiculously slow speed, chances are they are brand new on Morse code, so *send back* to them *at the same* exaggerated slower *speed*. **ANSWER B.**

G2C07 When sending CW, what does a "C" mean when added to the RST report?

A. Chirpy or unstable signal.
B. Report was read from S meter reading rather than estimated.
C. 100 percent copy.
D. Key clicks.

If someone sends you an RST report of 5-9-9 *C, it means your signal is unstable or chirping*, probably due to an inadequate power supply or operating from a low battery. **ANSWER A.**

Elmer Point: *The S meter on the front of your radio indicates signal strength. S-9 is much stronger than S-5, and S-1 is relatively weak, but it will be up to your own ears and brain to judge readability. R-2 means you can make out the signal with difficulty. R-5 is a loud and clear signal report. A great report would be 5 by 9. A bad one might be 3 by 3. Sometimes hams will generalize for strength, readability and CW tone as Q-5. Let's hope you always get a 5 by 9! Here's the how the RST Signal Reporting System works:*

The RST system is a way of reporting on the quality of a received signal by using a three digit number. The first digit indicates Readability (R), the second digit indicates received Signal Strength (S), and the third digit indicates Tone (T).

READABILITY (R) for Voice + CW
1 – Unreadable
2 – Barely readable, occasional words distinguishable
3 – Readable with considerable difficulty
4 – Readable with practically no difficulty
5 – Perfectly readable

SIGNAL STRENGTH (S) for Voice + CW
1 – Faint, barely perceptible signals
2 – Very weak signals
3 – Weak signals
4 – Fair signals
5 – Fairly good signals
6 – Good signals
7 – Moderately strong signals
8 – Strong signals
9 – Extremely strong signals

TONE* (T) Use on CW only
1 – Very rough, broad signals, 60 cycle AC may be present
2 – Very rough AC tone, harsh, broad
3 – Rough, low-pitched AC tone, some trace of filtering
5 – Filtered, rectified AC note, musical, ripple modulated
6 – Slight trace of filtered tone but with ripple modulation
7 – Near DC tone but trace of ripple modulation
8 – Good DC tone, may have slight trace of modulation
9 – Purest, perfect DC tone with no trace of ripple or modulation

*The TONE report refers only to the purity of the signal, and has no connection with its stability or freedom from clicks or chirps. If the signal has the characteristic steadiness of crystal control, add X to the report (e.g., RST 469X). If it has a chirp or "tail" (either on "make" or "break") add C (e.g., RST 469C). If it has clicks or other noticeable keying transients, add K (e.g., 469K). If a signal has both chirps and clicks, add both C and K (e.g., 469CK).

POPULAR Q SIGNALS

Given below are a number of Q signals whose meanings most often need to be expressed with brevity and clarity in amateur work. (Q abbreviations take the form of questions only when each is sent followed by a question mark.)

QRG Will you tell me my exact frequency (or that of _____)? Your exact frequency (or that of _____) is _____ kHz.

QRH Does my frequency vary? Your frequency varies.

QRI How is the tone of my transmission? The tone of your transmission is _____ (1. Good; 2. Variable; 3. Bad).

QRJ Are you receiving me badly? I cannot receive you. Your signals are too weak.

QRK What is the intelligibility of my signals (or those of _____)? The intelligibility of your signals (or those of _____) is _____ (1. Bad; 2. Poor; 3. Fair; 4. Good; 5. Excellent).

QRL Are you busy? I am busy (or I am busy with _____). Please do not interfere.

QRM Is my transmission being interfered with? Your transmission is being interfered with _____ (1. Nil; 2. Slightly; 3. Moderately; 4. Severely; 5. Extremely).

QRN Are you troubled by static? I am troubled by static _____ (1-5 as under QRM).

QRO Shall I increase power? Increase power.

QRP Shall I decrease power? Decrease power.

QRQ Shall I send faster? Send faster (_____ WPM).

QRS Shall I send more slowly? Send more slowly (_____ WPM).

QRT Shall I stop sending? Stop sending.

QRU Have you anything for me? I have nothing for you.

QRV Are you ready? I am ready.

QRW Shall I inform _____ that you are calling on _____ kHz? Please inform _____ that I am calling on _____ kHz.

QRX When will you call me again? I will call you again at _____ hours (on _____ kHz).

QRY What is my turn? Your turn is numbered _____ .

QRZ Who is calling me? You are being called by _____ (on _____ kHz).

QSA What is the strength of my signals (or those of _____)? The strength of your signals (or those of _____) is _____ (1. Scarcely perceptible; 2. Weak; 3. Fairly good; 4. Good; 5. Very good).

QSB Are my signals fading? Your signals are fading.

QSD Is my keying defective? Your keying is defective.

QSG Shall I send _____ messages at a time? Send _____ messages at a time.

QSK Can you hear me between your signals and if so can I break in on your transmission? I can hear you between my signals; break in on my transmission.

QSL Can you acknowledge receipt? I am acknowledging receipt.

QSM Shall I repeat the last message which I sent you, or some previous message? Repeat the last message which you sent me [or message(s) number(s) _____].

QSN Did you hear me (or _____) on _____ kHz? I heard you (or _____) on _____ kHz.

QSO Can you communicate with _____ direct or by relay? I can communicate with _____ direct (or by relay through _____).

QSP Will you relay to _____ ? I will relay to _____ .

QST General call preceding a message addressed to all amateurs and ARRL members. This is in effect "CQ ARRL."

QSU Shall I send or reply on this frequency (or on _____ kHz)?

QSW Will you send on this frequency (or on _____ kHz)? I am going to send on this frequency (or on _____ kHz).

QSX Will you listen to _____ on _____ kHz? I am listening to _____ on _____ kHz.

QSY Shall I change to transmission on another frequency? Change to transmission on another frequency (or on _____ kHz).

QSZ Shall I send each word or group more than once? Send each word or group twice (or _____ times).

QTA Shall I cancel message number _____ ? Cancel message number _____ .

QTB Do you agree with my counting of words? I do not agree with your counting of words. I will repeat the first letter or digit of each word or group.

QTC How many messages have you to send? I have messages for you (or for _____).

QTH What is your location? My location is _____ .

QTR What is the correct time? The time is _____ .

Source: ARRL

CW Lives!

G2C02 What should you do if a CW station sends "QRS"?

A. Send slower.
B. Change frequency.
C. Increase your power.
D. Repeat everything twice.

If you are sending code to another station and they respond *"QRS,"* this means for you to please *send* at a *slower* rate. **ANSWER A.**

G2C10 What does the Q signal "QRQ" mean?

A. Slow down.
B. Send faster.
C. Zero beat my signal.
D. Quitting operation.

If you are sending CW to another station, and they send back *"QRQ,"* they wish you to *send faster* because they are good at copying code at a higher speed. **ANSWER B.**

G2C11 What does the Q signal "QRV" mean?

A. You are sending too fast.
B. There is interference on the frequency.
C. I am quitting for the day.
D. I am ready to receive messages.

The Q code *"QRV" means* that you are *ready*, with pen or computer at hand, *to receive* the incoming message. If you send "QRV?," you are asking, "are there any messages holding for my station?" **ANSWER D.**

G2C09 What does the Q signal "QSL" mean?

A. Send slower.
B. We have already confirmed by card.
C. I acknowledge receipt.
D. We have worked before.

The term *"QSL"* has several meanings in ham radio. On CW and sometimes voice, it means that the *other station has acknowledged your message*. And if the other station asks for a QSL, they are likely asking for you to send them a colorful QSL card with your call sign on the front and QSO details on the back. QSL cards are fun to collect and all General Class hams should have their own QSL card. Also, you can QSL a contact electronically at several QSL e-mail sites. If ever you receive a QSL card, you should always send one back! **ANSWER C.**

G2C03 What does it mean when a CW operator sends "KN" at the end of a transmission?

A. Listening for novice stations.
B. Operating full break-in.
C. Listening only for a specific station or stations.
D. Closing station now.

When a station turns the communication back to you, and sends *"KN"* at the end of their transmission, it *means that they're wishing for only you to respond* and all other stations to stand by. **ANSWER C.**

COMMON CW ABBREVIATIONS

AA	All after	NR	Number
AB	All before	NW	Now; I resume transmission
ABT	About	OB	Old boy
ADR	Address	OM	Old man
AGN	Again	OP-OPR	Operator
ANT	Antenna	OT	Old timer; old top
AR	End of message	PBL	Preable
BCI	Broadcast interference	PSE-PLS	Please
BK	Break; break me; break in	PWR	Power
BN	All between; been	PX	Press
B4	Before	R	Received as transmitted; are
C	Yes	RCD	Received
CFM	Confirm; I confirm	REF	Refer to; referring to; reference
CK	Check	RPT	Repeat; I repeat
CL	I am closing my station; call	SED	Said
CLD-CLG	Called; calling	SEZ	Says
CUD	Could	SIG	Signature; signal
CUL	See you later	SKED	Schedule
CUM	Come	SRI	Sorry
CW	Continuous Wave	SVC	Service; prefix to service message
DLD-DLVD	Delivered	TFC	Traffic
DX	Distance	TMW	Tomorrow
FB	Fine business; excellent	TNX	Thanks
GA	Go ahead (or resume sending)	TU	Thank you
GB	Good-by	TVI	Television interference
GBA	Give better address	TXT	Text
GE	Good evening	UR-URS	Your; you're; yours
GG	Going	VFO-	Variable-frequency oscillator
GM	Good morning	VY	Very
GN	Good night	WA	Word after
GND	Ground	WB	Word before
GUD	Good	WD-WDS	Word; words
HI	The telegraphic laugh; high	WKD-WKG	Worked; working
HR	Here; hear	WL	Well; will
HV	Have	WUD	Would
HW	How	WX	Weather
KN	Listening for specific station(s)	XMTR	Transmitter
LID	A poor operator	XTAL	Crystal
MILS	Milliamperes	XYL	Wife
MSG	Message; prefix to radiogram	YL	Young lady
N	No	73	Best regards
ND	Nothing doing	88	Love and kisses
NIL	Nothing; I have nothing for you		

G2C08 What prosign is sent to indicate the end of a formal message when using CW?

A. SK.	C. AR.
B. BK.	D. KN.

The *end of a CW formal message* usually includes *"AR."* This lets all operators know that the formal message has been completely sent. **ANSWER C.**

G2C04 What does it mean when a CW operator sends "CL" at the end of a transmission?

A. Keep frequency clear.
B. Operating full break-in.
C. Listening only for a specific station or stations.
D. Closing station.

When an operator sends *"CL"* or "SK" at the end of their transmission, it means that they are *closing down their station* and going off the air. **ANSWER D.**

G2C01 Which of the following describes full break-in telegraphy (QSK)?

A. Breaking stations send the Morse code prosign BK.
B. Automatic keyers are used to send Morse code instead of hand keys.
C. An operator must activate a manual send/receive switch before and after every transmission.
D. Transmitting stations can receive between code characters and elements.

Although the FCC has eliminated the requirement to know Morse code, learning the code is important to be a good ham. All high frequency transceivers have Morse code capabilities, and many also include an automatic, built-in code keyer. Just add paddles! You will start out using the VOX mode to control the transmitter to turn on when you start keying, and turn off after a few seconds of not-keying. This is called semi-break-in. And when you really get good at CW, you can set your transceiver to operate CW in the *QSK full break-in mode*. In between each dot and dash, the receiver will instantly tune in on what's happening as you are sending CW on the bands. If another station wishes to interrupt, you will *hear its signal between your dots and dashes*. **ANSWER D.**

G1B03 Which of the following is a purpose of a beacon station as identified in the FCC Rules?

A. Observation of propagation and reception.
B. Automatic identification of repeaters.
C. Transmission of bulletins of general interest to Amateur Radio licensees.
D. Identifying net frequencies.

Beacon stations are important for the study of *propagation and reception* from the ionosphere. Always try to stay clear of beacon stations when selecting a frequency on which to transmit. Beacon stations are found at 14.100 MHz, 28.200 - 28.300 MHz, and up on the 2-meter band below 144.300 MHz. These are one-way transmissions. [97.3(a)(9)] **ANSWER A.**

G1B10 What is the power limit for beacon stations?

A. 10 watts PEP output.	C. 100 watts PEP output
B. 20 watts PEP output.	D. 200 watts PEP output

Beacon stations used for propagation surveys must never transmit more than *100 watts* peak envelope power *(PEP) output*. [97.203(c)] **ANSWER C.**

G1B02 With which of the following conditions must beacon stations comply?
A. A beacon station may not use automatic control.
B. The frequency must be coordinated with the National Beacon Organization.
C. The frequency must be posted on the Internet or published in a national periodical.
D. There must be no more than one beacon signal in the same band from a single location.

Beacon stations are important for the study of propagation from the ionosphere as well as the atmosphere. Do not transmit in beacon band plan frequencies. Ham operators are permitted to put up only a *single beacon signal, in the same band, from a single location*. [97.203(b)] **ANSWER D.**

Radio Beacon Stations

Slot	Country	Call	14.100	18.110	21.150	24.930	28.200	Operator
1	United Nations	4U1UN	00:00	00:10	00:20	00:30	00:40	UNRC
2	Canada	VE8AT	00:10	00:20	00:30	00:40	00:50	RAC
3	USA	W6WX	00:20	00:30	00:40	00:50	01:00	NCDXF
4	Hawaii	KH6WO	00:30	00:40	00:50	01:00	01:10	UHRO
5	New Zealand	ZL	00:40	00:50	01:00	01:10	01:20	NZART
6	Australia	VK8	00:50	01:00	01:10	01:20	01:30	W1A
7	Japan	JA21CY	01:00	01:10	01:20	01:30	01:40	JARL
8	China	BY	01:10	01:20	01:30	01:40	01:50	CRSA
9	Russia	UA	01:20	01:30	01:40	01:50	02:00	TBO
10	Sri Lanka	4S7B	01:30	01:40	01:50	02:00	02:10	RSSL
11	South Africa	ZS6DN	01:40	01:50	02:00	02:10	02:20	ZS6DN
12	Kenya	5Z4B	01:50	02:00	02:10	02:20	02:30	RSK
13	Israel	4X6TU	02:00	02:10	02:20	02:30	02:40	U of Tel Aviv
14	Finland	OH2B	02:10	02:20	02:30	02:40	02:50	U oh Helsinki
15	Madeira	CS3B	02:20	02:30	02:40	02:50	00:00	ARRM
16	Argentina	LU4AA	02:30	02:40	02:50	00:00	00:10	RCA
17	Peru	OA4B	02:40	02:50	00:00	00:10	00:20	RCP
18	Venezuela	YV5B	02:50	00:00	00:10	00:20	00:30	RCV

The 10-second, phase-3, message format is: "W6WX dah-dah-dah-dah" — each "dah" lasts a little more than one second. W6WX is transmitted at 100 watts, then each "dah" is attenuated in order, beginning at 100 watts, then 10 watts, then 1 watt, and finally 0.1 watt. *Courtesy CQ Magazine*

G4A02 What is one advantage of selecting the opposite or "reverse" sideband when receiving CW signals on a typical HF transceiver?
A. Interference from impulse noise will be eliminated.
B. More stations can be accommodated within a given signal passband.
C. It may be possible to reduce or eliminate interference from other signals.
D. Accidental out of band operation can be prevented.

I encourage you to learn Morse code. In the back of this book is a complete chapter on learning the code. On-air code practice is a great way to boost your copying speed. Sometimes when receiving a weak CW signal, you are hearing interference, too. Try this trick. *Switch* between upper and lower *sideband* to see if one of the sideband filters pulls in the CW signal better and *reduces the interference* from other signals. That's right – you can still hear CW in the SSB mode. Just be sure to go back to the CW mode when you are ready to send some code. **ANSWER C.**

Elmer Hint: LEARN CW

In Chapter 5 of this book, I give you a fun way to learn CW, beginning with the more common letters first. I have an 8 audio CD set for learning CW, too. Call (800) 669-9594 to learn more and order a copy for yourself.

You also can learn CW on-line! You can even track your progress on your computer. Visit these websites:
http://www.lcwo.net
http://aa9pw.com/morsecode
http://www.dxzone.com/catalog/Operating_Modes/Morse_code/Learning_Morse_Code

G8B09 Why is it good to match receiver bandwidth to the bandwidth of the operating mode?
A. It is required by FCC rules.
B. It minimizes power consumption in the receiver.
C. It improves impedance matching of the antenna.
D. It results in the best signal to noise ratio.
When you switch modes on your new radio, appropriate filters will fall into place giving you the *best signal-to-noise ratio*. You might be able to select tighter bandwidths, and this will further improve signal-to-noise ratios. **ANSWER D.**

G4A10 What is the purpose of an electronic keyer?
A. Automatic transmit/receive switching.
B. Automatic generation of strings of dots and dashes for CW operation.
C. VOX operation.
D. Computer interface for PSK and RTTY operation.
That brand new radio you rewarded yourself with in preparation for successfully passing your General exam will likely have a built-in *electronic keyer*. Sure, you could set the keyer to accept your granddad's old J-38 straight key, but why not try your CW skills with paddles, where your thumb and finger create *strings of dits and dahs*? **ANSWER B.**

Elmer Point: *Which is faster – new-fangled text messaging or old-reliable Morse code? Jay Leno wanted the answer to that question, so on May 13, 2006, Leno invited world text-messaging speed champ Ben Cook of Utah and his friend Jason to appear on* The Tonight Show with *Jay Leno to test their ability against Chip Margelli, K7JA and Ken Miller, K6CTW. Cook told Leno that he'd managed to send a 160-letter message to his friend in 57 seconds. Who won? Chip and Ken, hands down! Margelli sent his message at 29-wpm, but he was once timed sending code at 61.5 words per minute!*

Digital Operating

G2E07 What does the abbreviation "RTTY" stand for?
- A. Returning to you.
- B. Radioteletype.
- C. A general call to all digital stations.
- D. Repeater transmission type.

Although *radioteletype*, abbreviated *RTTY*, is a NO FRILLS type of communication on worldwide frequencies, there is still some RTTY traffic that you might want to tune in and decode. At night, tune in RTTY on 80 meters near 3600 kHz. During the day, try 15 meters around 21.080 kHz, and 20 meters at 14.080. **ANSWER B.**

G2E04 What segment of the 20 meter band is most often used for data transmissions?
- A. 14.000 - 14.050 MHz.
- B. 14.070 - 14.100 MHz.
- C. 14.150 - 14.225 MHz.
- D. 14.275 - 14.350 MHz.

Most RTTY transmissions on the *20 meter band* are found between *14.070 to 14.100 MHz*, a 30-kHz RTTY "window." Use the LSB switch to tune in 20 meter RTTY stations using a digital decoder. **ANSWER B.**

G2E05 Which of the following describes Baudot code?
- A. A 7-bit code, with start, stop and parity bits.
- B. A code using error detection and correction.
- C. A 5-bit code, with additional start and stop bits.
- D. A code using SELCAL and LISTEN.

Give me FIVE! This is a good way to remember *Baudot is a 5 bit code* with an additional start and stop bit. Give me FIVE! **ANSWER C.**

G2E06 What is the most common frequency shift for RTTY emissions in the amateur HF bands?
- A. 85 Hz.
- B. 170 Hz.
- C. 425 Hz.
- D. 850 Hz.

Almost all *ham radio RTTY* transmissions use the *170-Hz shift*. Commercial broadcast stations on shortwave use larger frequency shifts. **ANSWER B.**

G2E01 Which mode is normally used when sending an RTTY signal via AFSK with an SSB transmitter?

A. USB.   C. CW.
B. DSB.   D. LSB.

If you plan to operate radio teleprinter (*RTTY*) using audio-frequency-shift-keying (*AFSK*), your single sideband transmitter must be switched to *LSB, lower sideband*. **ANSWER D.**

☞ Visit: www.aorusa.com

You can connect a PSK-31 and RTTY data reader to your radio to decode messages.

G2E09 In what segment of the 20 meter band are most PSK31 operations commonly found?

A. At the bottom of the slow-scan TV segment, near 14.230 MHz.
B. At the top of the SSB phone segment near 14.325 MHz.
C. In the middle of the CW segment, near 14.100 MHz.
D. Below the RTTY segment, near 14.070 MHz.

Plenty of "whistles" heard are around *14.070 MHz on the 20 meter band*. Listen carefully – do you hear the slight warble? This is the sound of Phase Shift Keying (*PSK*), and with relatively inexpensive software you can easily decode this whistle into meaningful text scrolling across your screen. **ANSWER D.**

G2E02 How many data bits are sent in a single PSK31 character?

A. The number varies.   C. 7.
B. 5.   D. 8.

The sounds of *PSK31* take on the characteristics of a steady whistle on the airwaves with just a little warble. It's that variable warble that is part of VARICODE characters represented by *a variable-length combination of bits*. Just like the name VARICODE implies, the number of data bits VARIES. Good reading is the ARRL's *HF Digital Handbook*, edited by Steve Ford, WB8IMY. **ANSWER A.**

G8B10 What does the number 31 represent in PSK31?

A. The approximate transmitted symbol rate.
B. The version of the PSK protocol.
C. The year in which PSK31 was invented.
D. The number of characters that can be represented by PSK31.

Phase shift keying sounds like a steady carrier with a little superimposed warble on 14.070. *PSK31* only occupies about 31 Hertz of bandwidth, and the 31 also is the *approximate transmitted symbol rate*. There are some excellent software programs, many that you can download for free, to decode PSK31. **ANSWER A.**

G4A14 How should the transceiver audio input be adjusted when transmitting PSK31 data signals?
A. So that the transceiver is at maximum rated output power.
B. So that the transceiver ALC system does not activate.
C. So that the transceiver operates at no more than 25% of rated power.
D. So that the transceiver ALC indicator shows half scale.

When transmitting the solid carrier PSK31 data signal, make sure the automatic level control (*ALC*) indicator *doesn't wiggle at all*. Also, reduce power output to keep your transceiver from overheating. As long as your ALC meter is not flapping all over the place during PSK31 transmissions you probably have just the right amount of computer drive to your data controller box feeding the mic input.
ANSWER B.

Elmer Point: *Down on the digital frequencies you'll hear data activity, and with a simple sound card program you can decode all of the excitement. There are some regular digital contests, too, giving you a great way to begin transmitting and receiving data with other stations needing contact points. Here's a list of some of the bigger data contests:*

- New Years' Day RTTY Contest
- First weekend in January ARRL RTTY Roundup
- First weekend in February Low Power Digital Contest
- Second weekend in February Worldwide RTTY Call Letter Prefix Contest
- Second weekend in March RTTY Sprint
- Third weekend in April PSK-31 Activity weekend
- Third weekend in July North American RTTY Activity weekend
- First weekend in September PSK-31 Contest
- First weekend in October Hellschreiber Contest

G2E11 What does the abbreviation "MFSK" stand for?
A. Manual Frequency Shift Keying.
B. Multi (or Multiple) Frequency Shift Keying.
C. Manual Frequency Sideband Keying.
D. Multi (or Multiple) Frequency Sideband Keying .

The term *MFSK* stands for *Multi Frequency Shift Keying*, and the sound of the 16 tones on the air is quite distinctive. Its best quality is getting through when the band is fading out to a distant station. **ANSWER B.**

G2E10 What is a major advantage of MFSK16 compared to other digital modes?
A. It is much higher speed than RTTY.
B. It is much narrower bandwidth than most digital modes.
C. It has built-in error correction.
D. It offers good performance in weak signal environments without error correction.

MFSK16 sounds much like a calliope on the ham bands. Occupying just over 300 Hz of band width, MFSK16 is 16 baud with forward error correction. Tuning in an MFSK16 signal with the correct software requires spot-on accuracy, usually only found on a high frequency transceiver that can read out frequency to the Hz.

MFSK16 is an ideal mode (over RTTY) to overcome the problems of multipath. You will hear *MFSK* on the lower bands, like 80 meters, because it *will continue to get through when other digital modes fade out*. **ANSWER D.**

G8B12 What is the relationship between transmitted symbol rate and bandwidth?
A. Symbol rate and bandwidth are not related.
B. Higher symbol rates require higher bandwidth.
C. Lower symbol rates require higher bandwidth.
D. Bandwidth is constant for data mode signals.

You know why big boats go so slow out of the harbor, right? If they went any faster they would create big wakes that would upset every other boat on either side of them. Same thing for ham *radio digital modes* – if you're going to *send fast*, you'll need to go to higher bands that will permit the *faster sending speed and wider bandwidth*. **ANSWER B.**

Maximum Symbol (Baud) Rate for Amateur Bands

Amateur Band (meters)	Maximum Symbol Rate (bauds)
160 to 12 m	300 bauds
10 m	1200 bauds
6 and 2 m	19,600 bauds
1.25 and 0.70 m	56,000 bauds
33 cm and higher	Not Specified

G1C08 What is the maximum symbol rate permitted for RTTY or data emission transmitted at frequencies below 28 MHz?
A. 56 kilobaud.   C. 1200 baud.
B. 19.6 kilobaud.   D. 300 baud.

RTTY and data emissions below *28 MHz* (below the 10 meter band) must creep along no faster than *300 baud*. This slow symbol rate is required to minimize bandwidth allocation on the very crowded high-frequency bands. [97.307(f)(3)] **ANSWER D.**

G1C07 What is the maximum symbol rate permitted for RTTY or data emission transmission on the 20 meter band?
A. 56 kilobaud.   C. 1200 baud.
B. 19.6 kilobaud.   D. 300 baud.

Data emissions on the *20 meter band* must creep along no faster than *300 baud*. This slow symbol rate is required to minimize bandwidth usage on the very crowded high-frequency bands. [97.305(c), 97.307(f)(3)] **ANSWER D.**

G1C10 What is the maximum symbol rate permitted for RTTY or data emission transmissions on the 10 meter band?
A. 56 kilobaud.   C. 1200 baud.
B. 19.6 kilobaud.   D. 300 baud.

On *10 meters* we are permitted to increase RTTY and data emission transmission speeds up to *1200 baud*. This is a 4 times increase from the slower speed required on frequencies below the 10 meter band. [97.305(c) and 97.307(f)(4)] **ANSWER C.**

G1C11 What is the maximum symbol rate permitted for RTTY or data emission transmissions on the 2 meter band?

A. 56 kilobaud.　　　　　　　C. 1200 baud.
B. 19.6 kilobaud.　　　　　　D. 300 baud.

The next possible answer up from 1200 bauds is 19.6 kilobauds (kilo means 1000). *19,600 baud* is real quick for *2 meters*. [97.305(c) and 97.307(f)(5)] **ANSWER B.**

G1C09 What is the maximum symbol rate permitted for RTTY or data emission transmitted on the 1.25 meter and 70 centimeter bands?

A. 56 kilobaud.　　　　　　　C. 1200 baud.
B. 19.6 kilobaud.　　　　　　D. 300 baud.

We can open up the speed throttle when we transmit on *1.25 meters* (222 MHz band) and *70 cm* (440 MHz band) up to *56 kilobaud*. Here is where you will find some exciting data links. [97.305(c) and 97.307(f)(5)] **ANSWER A.**

G2E03 What part of a data packet contains the routing and handling information?

A. Directory.　　　　　　　　C. Header.
B. Preamble.　　　　　　　　D. Footer.

Within the *header* are *routing addresses* to digipeaters so your packet can be received and relayed over a specific route – even coast-to-coast and worldwide! If you are into APRS (Automatic Position/Packet Reporting System), you can set up your TNC (terminal node controller) header for local or wide-area relays of your position report. **ANSWER C.**

G2E08 What segment of the 80 meter band is most commonly used for data transmissions?

A. 3570 – 3600 kHz.　　　　C. 3700 – 3750 kHz.
B. 3500 – 3525 kHz.　　　　D. 3775 – 3825 kHz.

Down on the *80 meter band*, let your computer decode data. Tune from *3570 to 3600 kHz* for some fascinating data signals that only your computer can transform to a screen full of text. **ANSWER A.**

G6C10 What two devices in an Amateur Radio station might be connected using a USB interface?

A. Computer and transceiver.　　　C. Amplifier and antenna.
B. Microphone and transceiver.　　D. Power supply and amplifier.

The two devices at your new General Class station that could be *connected using a USB* interface are your *computer* and that brand new high frequency *transceiver*. **ANSWER A.**

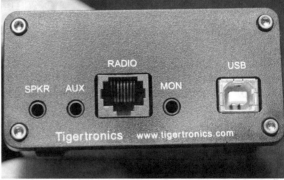

The universal serial bus (USB) has made it simple to connect your ham radio to your computer.

G6C12 Which of the following connectors would be a good choice for a serial data port?

A. PL-259.   C. Type SMA.
B. Type N.   D. DE-9.

Yup, answer D is spelled correctly, "DE-9." I know, I know, we usually call it a "DB-9." But for this examination go with the technically correct term "DE-9." If you have an older laptop and you want tie it in to your new ham radio to decode data, the computer may only offer a *DE-9 serial data port connector*. Most new data decoder programs now ship with a USB connector. So just remember, for this test the DE-9 connector is the correct answer. **ANSWER D.**

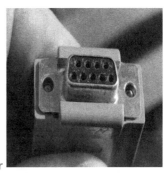

DE-9 connector

G6C14 Which of these connector types is commonly used for audio signals in Amateur Radio stations?

A. PL-259.   C. RCA Phono.
B. BNC.   D. Type N.

Let's look for the correct answer by identifying the use of each possible answer. A PL-259 is the coax cable connector that will join your new high frequency ham transceiver to your antenna. A BNC, along with the SMA, are antenna receptacles on hand-held transceivers. The Type N is an antenna connector for UHF and microwave frequencies. The *common connector for audio* frequencies is the *RCA phono jack*. **ANSWER C.**

RCA Connectors

G6C17 What is the general description of a DIN type connector?

A. A special connector for microwave interfacing.
B. A DC power connector rated for currents between 30 and 50 amperes.
C. A family of multiple circuit connectors suitable for audio and control signals.
D. A special watertight connector for use in marine applications.

DIN type connectors are found on the back of your HF transceiver, having up to 8 pin receptacles. Each manufacturer may have a unique arrangement of what is fed to the pin receptacles *used for audio and control signals*, so be careful when you start wiring in that new terminal node controller or linear amplifier relay and control. Sometimes there are small voltages on these receptacles, and you want to make absolutely sure you never short-out these voltage lines to ground because there may not be any fuse protection inside the radio's body. That's why I like the DIN plug already pre-soldered with wires. It's like doing brain surgery to try to solder up these connectors yourself. **ANSWER C.**

DIN connectors

G8B08 Why is it important to know the duty cycle of the data mode you are using when transmitting?
A. To aid in tuning your transmitter.
B. Some modes have high duty cycles which could exceed the transmitter's average power rating.
C. To allow time for the other station to break in during a transmission.
D. All of these choices are correct.

Hooking your high-frequency ham transceiver up to the modern computer opens up a whole new world of receiving and sending data signals. Receiving is almost a direct connection to your computer through the sound card, but sending data may require an external modem. Sending data with on-and-off handshake modes like PACTOR II or G-TOR cycles your new General Class transceiver to transmit and receive for a duty cycle (on transmit) of perhaps 50 percent. But other *modes like PSK31 and MFSK16* are a constant key-down data stream without interruption. Now your transmit *duty cycle is 100%*, and the back of your transceiver's heat sink is going to get roasty-toasty. To keep your equipment from going into meltdown in the data modes for transmitting, consider an external fan and reducing power output to the point the rear heat sinks won't fry eggs. If you get the heat sinks so warm that you can't touch them, you *could damage the transmitter* final output stage and that will lead to a mighty expensive repair. **ANSWER B.**

G2E13 In the PACTOR protocol, what is meant by an NAK response to a transmitted packet?
A. The receiver is requesting the packet be re-transmitted.
B. The receiver is reporting the packet was received without error.
C. The receiver is busy decoding the packet.
D. The entire file has been received correctly.

PACTOR is an exciting data mode that leads to near-perfect reception of a message. The transmitted message is sent in short bursts, allowing the receiving station to quickly transmit ACK for perfect copy, or *NAK*, meaning *"send it again, Sam, I missed a few characters."* This back and forth between two PACTOR stations leads to error-free copy. **ANSWER A.**

G8B11 How does forward error correction allow the receiver to correct errors in received data packets?
A. By controlling transmitter output power for optimum signal strength.
B. By using the varicode character set.
C. By transmitting redundant information with the data.
D. By using a parity bit with each character.

Forward error correction, abbreviated FEC, is achieved by *sending each character twice*. This allows the receiver to correct errors by double-checking the received data. This may be referred to as AMTOR (amateur teleprinting over radio) mode B. **ANSWER C.**

G2E12 How does the receiving station respond to an ARQ data mode packet containing errors?
A. Terminates the contact.
B. Requests the packet be retransmitted.
C. Sends the packet back to the transmitting station.
D. Requests a change in transmitting protocol.

ARQ stands for *Automatic Repeat Request*, and your Packet controller will request the sending station to *re-transmit the packet* if it detects errors in the reception of that data. **ANSWER B.**

G7C05 Which of the following is an advantage of a transceiver controlled by a direct digital synthesizer (DDS)?
A. Wide tuning range and no need for band switching.
B. Relatively high power output.
C. Relatively low power consumption.
D. Variable frequency with the stability of a crystal oscillator.

A *direct digital synthesizer* eliminates the need for multiple circuits of the local oscillator in your new high frequency transceiver. An HF rig with DDS may allow you to dial in a frequency all the way down to 1 Hz! If you plan to do a lot of digital work on the air, frequency stability and 1 Hz readout will be appreciated by the other operators. The entire DDS circuit is on one big chip, found in the phase locked loop board within your new HF transceiver. Initial concern was the generation of unwanted phase noise, but the design of modern HF transceivers all but eliminates this problem with direct digital synthesis. You can now *tune from 100 kHz through 54 MHz, with the stability of a temperature compensated crystal oscillator*. **ANSWER D.**

G7C11 What is meant by the term "software defined radio" (SDR)?
A. A radio in which most major signal processing functions are performed by software.
B. A radio which provides computer interface for automatic logging of band and frequency.
C. A radio which uses crystal filters designed using software.
D. A computer model which can simulate performance of a radio to aid in the design process.

You go over to your buddy's ham shack and you see his big computer connected to a tiny reading-book-sized box that contains the innards of the actual ham radio. Where's the rest of the gear? The *computer handles almost all the major signal processing* using software, and that little remaining box contains the final output transistors, and a cooling fan, to produce that great *software-defined radio* output on your favorite band. This is a super way to keep your rig up to date – just download any modifications, and presto, your rig is like brand new! **ANSWER A.**

DIGITAL ACCESSORIES

The following websites are excellent places for you to look for digital accessories for you ham radio and computer. Thanks to Don Wilson, N9ZGE, for sharing this with us.

AH-4	www.icomamerica.com/en/downloads/Default.aspx?Category=136
Airmail	www.siriuscyber.net/ham/
APRSPoint	www.aprspoint.com/
Digipan	www.digipan.net/
Donner	donnerstorenet.ipage.com/DCC/
Farallon Elecronics	www.farallon.us/webstore/
Ferrite Beads	www.dxengineering.com/Parts.asp?ID=1128&PLID=182&SecID=152& DeptID=42&PartNo=DXE-CSB-COMBO
Ferrite Beads	www.audiosystemsgroup.com/publish.htm
FTDI Drivers	www.ftdichip.com/FTDrivers.htm
GPS 18x OEM	buy.garmin.com/shop/shop.do?cID=158&pID=27594&ra=true
Ham Radio Deluxe	www.ham-radio-deluxe.com/
IC-718	www.icomamerica.com/en/products/amateur/hf/718/default.aspx
KPC3Plus	www.kantronics.com/products/kpc3.html
MFJ-1270C	www.mfjenterprises.com/man/pdf/MFJ-1270C.pdf
MicroHam	www.microham-usa.com/Products/USB3.html
MixW	http://mysite.verizon.net/jaffejim/
MMSSTV	http://mmhamsoft.amateur-radio.ca/pages/mmsstv.php
N4PY	www.n4py.com/
Packet Engine Pro	www.sv2agw.com/ham/pepro.htm
Palomar-Ferrite Cores	www.palomar-engineers.com/index.html
PrintScreen	www.gadwin.com/download/
PTC-IIusb	www.scs-ptc.com/shop/categories/modems-en
QuickMix	www.brothersoft.com/quickmix-10463.html
RIGblaster P&P	www.westmountainradio.com/product_info.php?products_id=pnp
RigExpert	www.rigexpert.com/index?s=standard
RMS Express & Paclink	www.winlink.org/ClientSoftware
SignaLink USB	www.tigertronics.com/
Sound Card Packet	www.kc2rlm.info/soundcardpacket/
TM-271	www.kenwoodusa.com/Communications/Amateur_Radio/ Mobiles/TM-271A
TM-271 Data Mod	www.kb2ljj.com/data/kenwood/TM-271A-E.htm
TM-271 Data Mod	www.mods.dk/view.php?ArticleId=3739
TNC-X	www.tnc-x.com/
Two Meter Antenna	www.mfjenterprises.com/Product.php?productid=MFJ-1750
USB to Serial Converter	www.ftdichip.com/Products/Cables/USBRS232.htm
Winlink Update	ftp://autoupdate.winlink.org/User%20Programs/

In An Emergency

G2B12 When is an amateur station allowed to use any means at its disposal to assist another station in distress?
  A. Only when transmitting in RACES.
  B. At any time when transmitting in an organized net.
  C. At any time during an actual emergency.
  D. Only on authorized HF frequencies.

One night you're tuning around the 20 meter band and you hear a maritime mobile ham calling for help on 14.150. Even though this is a frequency outside of your General Class limits, you are allowed to handle the distress call and possibly save some lives at sea! As long as the distress is a matter of life safety or the immediate protection of property, you are good to go with almost *any power level and any frequency*. But to do so, keep an accurate log, and *make certain that the communications qualify as an actual distress call*. [97.405(b)]
ANSWER C.

G2B11 What frequency should be used to send a distress call?
  A. Whatever frequency has the best chance of communicating the distress message.
  B. Only frequencies authorized for RACES or ARES stations.
  C. Only frequencies that are within your operating privileges.
  D. Only frequencies used by police, fire or emergency medical services.

Before you head out on your adventure, pre-plan what frequency is in use in your expected travel area that could hear your call for help. *Any active frequency would be a good spot to place a distress call*. Do NOT rely on known police, medical, or fire frequencies because most are protected with digital coded inputs that would not be able to decode your carrier. Rather, rely on normal, active, Amateur Radio communication frequencies. [97.405] **ANSWER A.**

G2B02 What is the first thing you should do if you are communicating with another amateur station and hear a station in distress break in?
  A. Continue your communication because you were on frequency first.
  B. Acknowledge the station in distress and determine what assistance may be needed.
  C. Change to a different frequency.
  D. Immediately cease all transmissions.

Act quickly to handle a station in distress. Find out WHO is in distress, WHERE they are located, and WHAT assistance may be needed. If it's a boat, find out how many persons are on board and instruct everyone to put on their personal floatation device. **ANSWER B.**

G1B04 Which of the following must be true before amateur stations may provide communications to broadcasters for dissemination to the public?

A. The communications must directly relate to the immediate safety of human life or protection of property and there must be no other means of communication reasonably available before or at the time of the event.

B. The communications must be approved by a local emergency preparedness official and conducted on officially designated frequencies.

C. The FCC must have declared a state of emergency.

D. All of these choices are correct.

Your new General Class ham station is not to be used for routine news gathering for your local television station. However, *in a disaster*, where a ham on scene tells you the center span of a bridge has just collapsed, it *WOULD BE permissible for you to contact your local news agency and relay this safety of human life transmission* so motorists don't accidentally fly off the span. [97.113(b)] **ANSWER A.**

G2B09 Who may be the control operator of an amateur station transmitting in RACES to assist relief operations during a disaster?

A. Only a person holding an FCC issued amateur operator license.

B. Only a RACES net control operator.

C. A person holding an FCC issued amateur operator license or an appropriate government official.

D. Any control operator when normal communication systems are operational.

Radio Amateur Civil Emergency Service (*RACES*) is a public service by *licensed amateur radio operators* to provide volunteer communications to government agencies in time of extraordinary need. During periods of RACES activation, licensed amateur radio operators may serve their local government agency. The Federal Emergency Management Agency (FEMA) provides planning guidance and technical assistance for establishing a RACES unit at the state and local government levels. Only licensed amateur radio operators with a current RACES authorization may be the control operator at a RACES station. [97.407(a)] **ANSWER A.**

In an emergency, authorized hams participating in a RACES organization may communicate from a police helicopter.

G2B10 When may the FCC restrict normal frequency operations of amateur stations participating in RACES?

A. When they declare a temporary state of communication emergency.
B. When they seize your equipment for use in disaster communications.
C. Only when all amateur stations are instructed to stop transmitting.
D. When the President's War Emergency Powers have been invoked.

During a time of war, when the President exercises his *War Emergency Powers*, *RACES* may become the ONLY communications allowed via amateur radio. All other non-RACES ham operators would be ordered to stop transmitting – only RACES operators could remain on the air. [97.407(b)] **ANSWER D.**

RACES Logo.

▼ IF YOU'RE LOOKING FOR	▼ THEN VISIT
All About RACES	www.USRACES.org
News on Emergency Groups	www.N4KSS.net/Reflectors.html
All about ARES	www.QSL.net/ARES
Military Radio Groups	www.NAVYMARS.org
More About Joining ARES	www.ARRL.org/ARES
ARRL Emergency Volunteers	www.ARRL.org/Volunteer

The Effect of the Ionosphere on Radio Waves

To help you with the questions on radio wave propagation, here is a brief explanation on the effect the ionosphere has on radio waves.

The ionosphere is the electrified atmosphere from 40 miles to 400 miles above the Earth. You can sometimes see it as "northern lights." It is charged-up daily by the Sun, and does some miraculous things to radio waves that strike it. Some radio waves are absorbed during daylight hours by the ionosphere's D layer. Others are bounced back to Earth. Yet others penetrate the ionosphere and never come back again. The wavelength of the radio waves determines whether the waves will be absorbed, refracted, or will penetrate. Here's a quick way to memorize what the different layers do during day and nighttime hours:

The D layer is about 40 miles up. The D layer is a Daylight layer; it almost disappears at night. D for Daylight. The D layer absorbs radio waves between 1 MHz to 7 MHz. These are long wavelengths. All others pass through.

The E layer is also a daylight layer, and it is very Eccentric. E for Eccentric. Patches of E layer ionization may cause some surprising reflections of signals on both high frequency as well as very-high frequency. The E layer height is usually 70 miles.

The F1 layer is one of the layers farthest away. The F layer gives us those Far away signals. F for Far away. The F1 layer is present during daylight hours, and is up around 150 miles. The F2 layer is also present during daylight hours, and it gives us the Furthest range. The F2 layer is 250 miles high, and it's best for the Farthest range on medium and short waves. The F2 layer is strongest in the summer months. During winter months, both the F1 and F2 layers may become unpredictable, but always strong enough to support exciting skywaves! At nighttime, the F1 and F2 layers combine to become just the F layer at 180 miles. This F layer at nighttime will usually bend radio waves between 1 MHz and 15 MHz back to earth. At night, the D and E layers disappear.

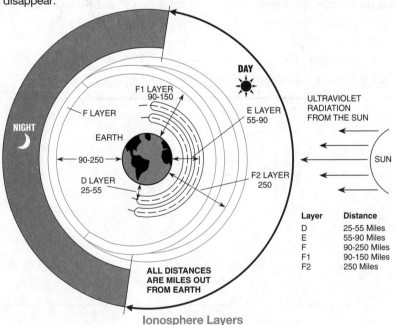

Ionosphere Layers

Source: *Antennas — Selection and Installation*, © 1986, Master Publishing, Inc., Niles, Illinois

Skywave Excitement

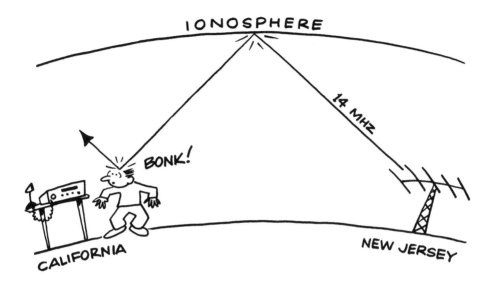

IONOSPHERE

14 MHz

BONK!

CALIFORNIA

NEW JERSEY

Elmer Point: *One of the most fascinating aspects of General Class medium frequency and high frequency operation is propagation. Your HF signals are going coast to coast, thanks to Mother Nature's ionosphere. Many new General Class operators, with a modest dipole antenna, are spellbound when a station from Europe or the Far East comes in out of the noise, builds to S-9, and gives you a solid signal report with your paltry 100 watts and unity gain dipole antenna strung up to an apple tree. And just as soon as they give you that great report, the signal begins to fade away. Welcome to propagation. This topic contains the MOST questions for your General Class studies, and this is perfect timing because this General Class question pool remains valid as we approach the peak of our next solar cycle. Propagation will keep you at the edge of your radio room chair when the band magically OPENS!*

G3C03 Why is the F2 region mainly responsible for the longest distance radio wave propagation?
A. Because it is the densest ionospheric layer.
B. Because it does not absorb radio waves as much as other ionospheric regions.
C. Because it is the highest ionospheric region.
D. All of these choices are correct.
The higher the altitude of the ionospheric region refracting the high-frequency radio waves, the greater the radio range. The *F2* layer is the *highest layer* and gives the *longest skywave propagation*. **ANSWER C.**

G3B09 What is the approximate maximum distance along the Earth's surface that is normally covered in one hop using the F2 region?
    A. 180 miles.   C. 2,500 miles.
    B. 1,200 miles.   D. 12,000 miles.

The *F2 layer* is our highest reflective ionospheric region, approximately 250 miles up. This gives worldwide signals their furthest bounce, generally about *2500 miles*. **ANSWER C.**

G3C02 Where on the Earth do ionospheric layers reach their maximum height?
    A. Where the Sun is overhead.
    B. Where the Sun is on the opposite side of the Earth.
    C. Where the Sun is rising.
    D. Where the Sun has just set.

Our ionosphere is influenced by ultraviolet radiation from the Sun. Maximum ultraviolet radiation occurs around noon, local time, *when the Sun is at its highest* elevation overhead, which is when our *ionospheric layers reach their maximum height*. **ANSWER A.**

G3C04 What does the term "critical angle" mean as used in radio wave propagation?
    A. The long path azimuth of a distant station.
    B. The short path azimuth of a distant station.
    C. The lowest takeoff angle that will return a radio wave to the Earth under specific ionospheric conditions.
    D. The highest takeoff angle that will return a radio wave to the Earth under specific ionospheric conditions.

Have you ever skipped stones on a lake? There is an angle that you cannot exceed where the stone doesn't skip, but rather penetrates into the water. Radio waves in the ionosphere act similarly; there is a point – the highest take-off angle – that cannot be exceeded or a radio wave will not reflect back to Earth. Just remember *"highest take-off angle."* **ANSWER D.**

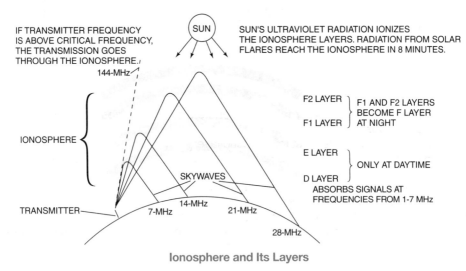

Ionosphere and Its Layers

G3B08 What does MUF stand for?

A. The Minimum Usable Frequency for communications between two points.
B. The Maximum Usable Frequency for communications between two points.
C. The Minimum Usable Frequency during a 24 hour period.
D. The Maximum Usable Frequency during a 24 hour period.

The *maximum usable frequency (MUF)* peaks during our day in the morning hours to the east, for working Europe, and in the afternoon and evening hours, to the west, working Asia. This is the fun of *working ham radio skywaves* – the ionosphere and the maximum usable frequency will constantly give us some excitement if we just keep tuned in. **ANSWER B.**

G3B05 What usually happens to radio waves with frequencies below the Maximum Usable Frequency (MUF) and above the Lowest Usable Frequency (LUF) when they are sent into the ionosphere?

A. They are bent back to the Earth.
B. They pass through the ionosphere.
C. They are amplified by interaction with the ionosphere.
D. They are bent and trapped in the ionosphere to circle the Earth.

Frequencies below the maximum usable frequency and above the Lowest Usable Frequency are *bent back to Earth* by the ionosphere. For maximum range, operate as close to MUF as possible. **ANSWER A.**

G3B04 What is a reliable way to determine if the Maximum Usable Frequency (MUF) is high enough to support skip propagation between your station and a distant location on frequencies between 14 and 30 MHz?

A. Listen for signals from an international beacon.
B. Send a series of dots on the band and listen for echoes from your signal.
C. Check the strength of TV signals from Western Europe.
D. Check the strength of signals in the MF AM broadcast band.

Is a particular band open to the rest of the world? A simple way to find out is spin the big VFO knob and see what you hear. If it is full of signals, the band is open to somewhere in the world. If you spin the knob and all you hear is a couple of local hams yakking back and forth, the absence of activity indicates the band is not propagating very far! I like to listen to the Northern California DX Foundation *beacons*, at 14.100 MHz. If you know a little bit of CW, and keep track of precise time, you'll know which part of the world is coming in on 20 meters and higher.

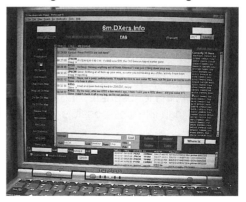

> 14.100 MHz
> 18.110 MHz
> 21.150 MHz
> 24.930 MHz
> 28.200 MHz

Each beacon transmits its CW call sign on a rotational basis, followed by four 1-second dashes. The call sign and first dash are sent at 100 watts, the second dash at 10 watts, the third dash at 1 watt. You're doing great if you can hear the last dash at 100 milliwatts! (visit www.NCDXF.org/beacon) **ANSWER A.**

There are websites that provide skywave DX conditions.

G3B03 Which of the following applies when selecting a frequency for lowest attenuation when transmitting on HF?
  A. Select a frequency just below the MUF.
  B. Select a frequency just above the LUF.
  C. Select a frequency just below the critical frequency.
  D. Select a frequency just above the critical frequency.

The term *"lowest attenuation"* means least amount of signal fading. The trick is to try to operate on the highest high frequency ham band that gives you skywave propagation to somewhere else in the country or the world. Operating *just below the maximum usable frequency (MUF)* will lead to some extraordinary crystal-clear communications. **ANSWER A.**

G3B12 What factors affect the Maximum Usable Frequency (MUF)?
  A. Path distance and location.
  B. Time of day and season.
  C. Solar radiation and ionospheric disturbances.
  D. All of these choices are correct.

Getting a signal halfway around the Earth, operating at just below the maximum usable frequency, is a ham radio tradition. But there are plenty of things to consider – solar activity, day or night, fall or summer, and how far away the other station is located. *All of these choices are correct.* **ANSWER D.**

G3B10 What is the approximate maximum distance along the Earth's surface that is normally covered in one hop using the E region?
  A. 180 miles.   C. 2,500 miles.
  B. 1,200 miles.   D. 12,000 miles.

The *E layer* is between 50 and 90 miles up and, because it's closer to Earth, high-frequency waves don't bounce as far as they do off of the F layer. E skip is approximately *1200 miles* and, during the summertime, "sporadic E" may sometimes "short skip" in as close as 600 miles. **ANSWER B.**

G3B02 Which of the following is a good indicator of the possibility of skywave propagation on the 6 meter band?
  A. Short skip skywave propagation on the 10 meter band.
  B. Long skip skywave propagation on the 10 meter band.
  C. Severe attenuation of signals on the 10 meter band.
  D. Long delayed echoes on the 10 meter band.

Now that you are upgrading from Technician to General, don't abandon all the excitement on 6 meters. In fact, your new General Class privileges on 10 meters will help you forecast when we might have an upcoming 6-meter band opening. When you begin to make contact via *skywaves on 10 meters with stations less than 300 miles away*, this indicates an extremely strong Sporadic E-skip "cloud" out there in the ionosphere, and *E-skip on 6 meters* and maybe even 2 meters will more than likely be possible within the next half hour! **ANSWER A.**

G3C01 Which of the following ionospheric layers is closest to the surface of the Earth?

A. The D layer.   C. The F1 layer.
B. The E layer.   D. The F2 layer.

The Darn *D layer*! It is the layer *closest to* the surface of the *Earth*, and during daylight hours it is usually responsible for absorbing ham radio medium-frequency skywave signals. **ANSWER A.**

Altitudes in Miles of Ionospheric Layers

| | Day | | |
Layers	Summer	Winter	Night
F2	>250		
F1	90-150		
F		90-150	90-250
E	55-90	55-90	
D	40	40	

G3C12 Which ionospheric layer is the most absorbent of long skip signals during daylight hours on frequencies below 10 MHz?

A. The F2 layer.   C. The E layer.
B. The F1 layer.   D. The D layer.

The 40-meter band is a great one for 500-mile, daylight, skywave contacts. Even though they may fade in and out a little bit, they are almost always there from Sun-up to around 4:00 p.m. local time. After 4:00 p.m., the D-layer associated with absorption begins to disappear, and the 40 meter band begins to go "long." Your 500-mile buddies will disappear, and next thing you hear are stations a couple thousand miles away pouring in. Then, as the 40 meter band continues on well into the night, in come the dreaded megawatt, foreign, double-sideband, shortwave broadcast stations that share our frequencies, too. What you will hear at night and in the early morning hours on 40 meters are extremely loud whistles from foreign broadcast, double-sideband, full-carrier stations, and the key to operating 40 meters in the early morning hours is finding a spot to dodge the foreign broadcast heterodynes. But during the day, the Darn *D Layer causes fades in local 500-mile contacts*. **ANSWER D.**

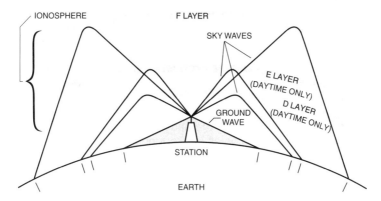

Radio Wave Propagation
Source: *Mobile 2-Way Radio Communications,* G. West, © 1993, Master Publishing, Inc.

G3C05 Why is long distance communication on the 40, 60, 80 and 160 meter bands more difficult during the day?
  A. The F layer absorbs signals at these frequencies during daylight hours.
  B. The F layer is unstable during daylight hours.
  C. The D layer absorbs signals at these frequencies during daylight hours.
  D. The E layer is unstable during daylight hours.

The "Darn D" layer does more harm than good to medium- and high-frequency signals. During daylight hours, SSB and CW operation on 160 and 80 meters is confined to ground wave coverage. On 40 meters, daytime skip distances are generally no greater than 600 miles when the *D layer* is doing its thing *absorbing MF and HF signals*. **ANSWER C.**

G3B07 What does LUF stand for?
  A. The Lowest Usable Frequency for communications between two points.
  B. The Longest Universal Function for communications between two points.
  C. The Lowest Usable Frequency during a 24 hour period.
  D. The Longest Universal Function during a 24 hour period.

The term *"lowest usable frequency"* refers specifically to *skywave stations attempting contact*. On high frequency ground waves, out to 20 miles, you will get through no matter what the ionosphere is doing. But if you and a buddy want to stay in touch on the 75 meter ham band, separated by 500 miles, you may find early morning contacts loud and clear, but at high noon, the lowest usable frequency is many megahertz above where you plan to operate, and no contact will be made. But hang around – the LUF will begin to drop quickly around sundown. **ANSWER A.**

G3B06 What usually happens to radio waves with frequencies below the Lowest Usable Frequency (LUF)?
  A. They are bent back to the Earth.
  B. They pass through the ionosphere.
  C. They are completely absorbed by the ionosphere.
  D. They are bent and trapped in the ionosphere to circle the Earth.

During daylight hours, the lowest usable frequency for a skywave contact may be around 3 MHz. This means the 160 meter band, just under 2 MHz, will not be usable for daylight *skywave* contacts because they are *completely absorbed by the ionosphere*. But as soon as the Sun goes down, hang on for some great DX contacts! **ANSWER C.**

G3B11 What happens to HF propagation when the Lowest Usable Frequency (LUF) exceeds the Maximum Usable Frequency (MUF)?
  A. No HF radio frequency will support ordinary skywave communications over the path.
  B. HF communications over the path are enhanced.
  C. Double hop propagation along the path is more common.
  D. Propagation over the path on all HF frequencies is enhanced.

When the lowest usable frequency (LUF) jumps up and actually exceeds the maximum usable frequency (MUF), high frequency *radio communications along a specific ionospheric signal path will disappear*. This sometimes occurs with increased geomagnetic activity from Sunspots. **ANSWER A.**

G3C09 What type of radio wave propagation allows a signal to be detected at a distance too far for ground wave propagation but too near for normal skywave propagation?

A. Faraday rotation.

B. Scatter.

C. Sporadic-E skip.

D. Short-path skip.

Backscatter communications is one way to reach a station that is in that zone of no-reception – the skip zone. When I communicate from southern California to Seattle, San Francisco is in my skip zone and will not receive my signals. But if I aim my beam antenna west toward Hawaii, some of my signal is backscattered into the Bay Area, giving me communications to a station that is too far for ground wave, and too close for normal skywaves. **ANSWER B.**

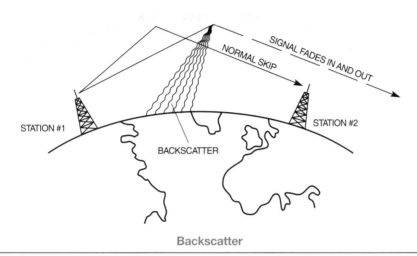

Backscatter

G3C08 Why are HF scatter signals in the skip zone usually weak?

A. Only a small part of the signal energy is scattered into the skip zone.

B. Signals are scattered from the magnetosphere which is not a good reflector.

C. Propagation is through ground waves which absorb most of the signal energy.

D. Propagation is through ducts in the F region which absorb most of the energy.

During periods of scatter communications, *only a fraction of the original signal is scattered* back to those stations too far for ground wave reception, yet too close for the main part of your signal being reflected by the ionosphere. Some of the scattered signals are refracted back into your skip zone, so all you get is a very weak incoming skywave. **ANSWER A.**

G3C06 What is a characteristic of HF scatter signals?

A. They have high intelligibility.

B. They have a wavering sound.

C. They have very large swings in signal strength.

D. All of these choices are correct.

High frequency *scatter communications* bounce a portion of your signal off of densely ionized patches in the ionosphere. Since the ionosphere is constantly in motion, the signals will fade in and out, much like ocean waves, resulting in a *wavering sound*. **ANSWER B.**

G3C07 What makes HF scatter signals often sound distorted?
A. The ionospheric layer involved is unstable.
B. Ground waves are absorbing much of the signal.
C. The E-region is not present.
D. Energy is scattered into the skip zone through several different radio wave paths.

The wavy sound of HF scatter signals, especially backscatter, is caused by the signal being reflected back through several radio wave paths, creating *multi-path distortion*. **ANSWER D.**

G3C10 Which of the following might be an indication that signals heard on the HF bands are being received via scatter propagation?
A. The communication is during a sunspot maximum.
B. The communication is during a sudden ionospheric disturbance.
C. The signal is heard on a frequency below the Maximum Usable Frequency.
D. The signal is heard on a frequency above the Maximum Usable Frequency.

If you chose a *frequency slightly above the maximum usable frequency*, you can sometimes take advantage of the ionosphere to scatter your communications to an area that normally would not hear radio wave reflection from the ionosphere. **ANSWER D.**

G2D04 Which of the following describes an azimuthal projection map?
A. A world map that shows accurate land masses.
B. A world map projection centered on a particular location.
C. A world map that shows the angle at which an amateur satellite crosses the equator.
D. A world map that shows the number of degrees longitude that an amateur satellite appears to move westward at the equator with each orbit.

Long-range communications do not necessarily go in straight lines. When we navigate our signals around the world, we need a *chart* that takes into account the curvature of the Earth and *where we are located* upon it. An azimuthal map shows your ham shack at the center of the Earth and will help you determine the shortest path between your station and that rare DX station. **ANSWER B.**

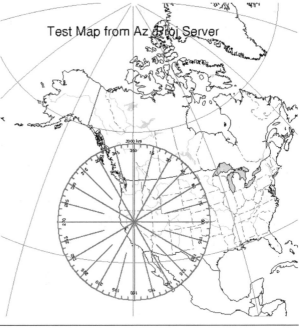

Test Map from Az Proj Server

Ham operators with a beam antenna will use an azimuthal map like this one to determine short path and long path headings to reach DX stations.

G3B01 How might a skywave signal sound if it arrives at your receiver by both short path and long path propagation?

A. Periodic fading approximately every 10 seconds.
B. Signal strength increased by 3 dB.
C. The signal might be cancelled causing severe attenuation.
D. A well-defined echo might be heard.

Do you have my audio course? If so, you can hear the distinctive sound of simultaneous short-path and long-path reception. Incoming *signals will have a well-defined echo* because the short-path signal is coming to you from a much shorter distance than the long-path signal coming all the way around the globe. **ANSWER D.**

G2D06 How is a directional antenna pointed when making a "long-path" contact with another station?

A. Toward the rising Sun.
B. Along the gray line.
C. 180 degrees from its short-path heading.
D. Toward the north.

Some ionospheric conditions may allow you to establish communications with a distant station on General Class worldwide frequencies over a longer path around the world than the direct short path. If you hear the station with an echo, try turning your beam *antenna 180 degrees* in the opposite direction *from the short pat*h direction to see whether or not the station will come in better on the long path. The echo you hear is caused by the difference in the time it takes the signal to reach you between long path and short path. **ANSWER C.**

G3A11 Approximately how long is the typical sunspot cycle?

A. 8 minutes.   C. 28 days.
B. 40 hours.   D. 11 years.

Our *sunspot cycle* peaks every *11 years*. We are now heading for the peak of solar cycle 24, expected to climb to its highest activity around 2015 or 2016. This is good timing for your new General Class privileges – as we near the peak, extraordinary long-range skywave conditions will prevail on the 10 to 20 meter bands, with 20 meters staying "open" almost all night long at the peak! **ANSWER D.**

G3A01 What is the sunspot number?

A. A measure of solar activity based on counting sunspots and sunspot groups.
B. A 3 digit identifier which is used to track individual sunspots.
C. A measure of the radio flux from the Sun measured at 10.7 cm.
D. A measure of the sunspot count based on radio flux measurements.

We have been studying the surface of the Sun since 1610. The Zurich Observatory was built in 1749, and 100 years later began to regularly record Sunspot numbers. Every 11 years, the Sun goes through a period of huge solar eruptions of charged particles. These charged electrons and ions from the Sun become a solar wind that streams into our upper magnetosphere, which sometimes enhances ham radio transmissions on VHF bands, and sometimes soaks-up normal skywave contacts on high frequency. A report of the sunspot number refers to a *daily index of sunspot activity*. **ANSWER A.**

G3A10 What causes HF propagation conditions to vary periodically in a 28-day cycle?

A. Long term oscillations in the upper atmosphere.
B. Cyclic variation in the Earth's radiation belts.
C. The Sun's rotation on its axis.
D. The position of the Moon in its orbit.

We find recurring skywave conditions about every 28 days, as the *Sun makes a complete rotation*. New satellites continuously stare at the Sun, even giving glimpses of solar eruptions soon to take aim at Earth. Ham operators carefully monitor solar activity, and find that stronger sunspots may reappear 28 days later for more excitement on the air waves. Visit www.spaceweather.com **ANSWER C.**

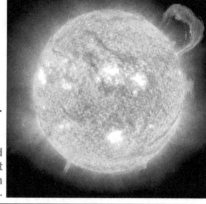

Solar flares and sunspots affect radiowave propagation
Photo courtesy of N.A.S.A.

G3A09 What effect do high sunspot numbers have on radio communications?

A. High-frequency radio signals become weak and distorted.
B. Frequencies above 300 MHz become usable for long-distance communication.
C. Long-distance communication in the upper HF and lower VHF range is enhanced.
D. Microwave communications become unstable.

We are just beginning an upward climb of a new solar cycle, cycle 24. We can expect to see the cycle peak around 2015, and plan for *fun operation on high frequencies* throughout the entire solar cycle climb. Sunspots will become more common as we approach this solar cycle peak in a few more years. **ANSWER C.**

G3A04 Which of the following amateur radio HF frequencies are least reliable for long distance communications during periods of low solar activity?

A. 3.5 MHz and lower.   C. 10 MHz.
B. 7 MHz.   D. 21 MHz and higher.

A couple of years ago, we were at the bottom of solar cycle 23. When that cycle ended, there was nearly a year *without* ANY *sunspots*, and frequencies from *21 MHz (15 meters) and higher (10 meters) were absent of daily skywaves*. But we're past the doldrums! Hurray! Sunspots are now a daily occurrence, so you are earning your new General Class license at just the right time for increased solar – and skywave – activity! **ANSWER D.**

G3A05 What is the solar-flux index?

A. A measure of the highest frequency that is useful for ionospheric propagation between two points on the Earth.
B. A count of sunspots which is adjusted for solar emissions.
C. Another name for the American sunspot number.
D. A measure of solar radiation at 10.7 cm.

The *solar-flux* index is *measured* daily on *10.7 cm* in Ottawa, Canada. You may tune into the radio propagation solar activity reports transmitted by WWV at 18 minutes past the hour. Frequencies of 10 and 15 MHz will give you best reception during the day, and 5 MHz may give you best reception at night. **ANSWER D.**

Elmer Point: *Your new HF General Class radio also offers full shortwave reception. Try 10- or 15-MHz to hear the WWV time ticks. At 18 minutes past the hour, listen to your latest ionospheric and solar weather reports! A K index between 1 to 4, and an A index of 0 to 7 means the ionosphere is quiet and very predictable for long-range skywaves. But a K index of 5 or 6 with an A index of 30 to 49 indicates that HF band conditions will become unpredictable and unstable. And hang on to your receivers – a K index of 7 to 9 or an A index of 50 and higher means the HF bands will likely be unusable for regular skywave contacts due to a major solar storm with strong major geomagnetic activity disrupting HF signals. With high K and A indexes, time to go outside after dark and look for an aurora! Here's a summary of K and A Index readings:*

K Index	A Index	HF Skip Conditions
K1 – K4	A0 – A7	Bands are normal
K4	A8 – A15	Bands are unsettled
K4	A16 – A30	Bands are unpredictable
K5	A30 – A50	Lower bands are unstable
K6	A50 – A99	Few skywaves below 15 MHz
K7 – K9	A100 – A400	Radio blackout is likely.
		Go fishing or watch for an aurora.

G3A12 What does the K-index indicate?
A. The relative position of sunspots on the surface of the Sun.
B. The short term stability of the Earth's magnetic field.
C. The stability of the Sun's magnetic field.
D. The solar radio flux at Boulder, Colorado.

Earth is surrounded by the magnetosphere that acts as a barrier protecting us from some of the charged particles coming from eruptions on the Sun. The magnetosphere occasionally develops holes and cracks that may allow tremendous amounts of solar energy to disturb our natural *geomagnetic stability*. Some of this energy that penetrates the magnetosphere can be seen as auroras. A low *K index* means good, stable high-frequency propagation. **ANSWER B.**

G3A13 What does the A-index indicate?
A. The relative position of sunspots on the surface of the Sun.
B. The amount of polarization of the Sun's electric field.
C. The long term stability of the Earth's geomagnetic field.
D. The solar radio flux at Boulder, Colorado.

The *A-Index* is a 24 hour averaging of the planetary K-Index, and is a great indicator of *long term stability* of the Earth's geomagnetic field. **ANSWER C.**

G3A03 Approximately how long does it take the increased ultraviolet and X-ray radiation from solar flares to affect radio-wave propagation on the Earth?
    A. 28 days.   C. 8 minutes.
    B. 1 to 2 hours.   D. 20 to 40 hours.
Ultraviolet radiation travels at the speed of light. It takes about *8 minutes* for sunlight and ultraviolet rays to reach the Earth's ionosphere. Sunspots may quickly appear, and heavy sunspot activity may affect worldwide propagation for up to 3 days. **ANSWER C.**

G3A06 What is a geomagnetic storm?
    A. A sudden drop in the solar-flux index.
    B. A thunderstorm which affects radio propagation.
    C. Ripples in the ionosphere.
    D. A temporary disturbance in the Earth's magnetosphere.
We're surrounded! Above our weather layer of the atmosphere, called the troposphere, is the ionosphere with its D layer, and above that the E layer, and on top of the E layer are the F-1 and F-2 layers. But did you know there is yet another layer up around 400 miles, called the magnetosphere? When the *Sun sends out a coronal mass ejection*, the charged particles may *interact with our magnetosphere*, and disrupt high frequency communications for several days. **ANSWER D.**

G3A02 What effect does a Sudden Ionospheric Disturbance have on the daytime ionospheric propagation of HF radio waves?
    A. It enhances propagation on all HF frequencies.
    B. It disrupts signals on lower frequencies more than those on higher frequencies.
    C. It disrupts communications via satellite more than direct communications.
    D. None, because only areas on the night side of the Earth are affected.
The *lower bands*, such as 160, 80, 40, and even 20 meters, *become so noisy* that it is impossible to hear any distant signals coming in from skywaves during a sudden ionospheric disturbance. **ANSWER B.**

G3A08 Which of the following effects can a geomagnetic storm have on radio-wave propagation?
    A. Improved high-latitude HF propagation.
    B. Degraded high-latitude HF propagation.
    C. Improved ground-wave propagation.
    D. Improved chances of UHF ducting.
During periods of major geomagnetic disturbances, high-frequency propagation over *high-latitude paths will be degraded*. 6 meters VHF might be hopping! **ANSWER B.**

G3A16 What is a possible benefit to radio communications resulting from periods of high geomagnetic activity?
    A. Aurora that can reflect VHF signals.
    B. Higher signal strength for HF signals passing through the polar regions.
    C. Improved HF long path propagation.
    D. Reduced long delayed echoes.
During periods of high geomagnetic activity, beautiful auroras may be viewed at

night, away from city lights, as far south as latitude 37° north. The aurora itself is created when gases in our upper atmosphere are bombarded by coronal mass ejections, kind of like a neon light. Depending on which gas gets stirred up determines what color Northern Lights you may see. A trip to Alaska during equinox will usually lead to some awe-inspiring Northern Lights viewing. You will definitely be impressed by an aurora. *The aurora can also reflect VHF signals*, too, adding to the excitement. **ANSWER A.**

Geomagnetic disturbances caused by the Sun result in the Northern Lights.

G3A15 How long does it take charged particles from coronal mass ejections to affect radio-wave propagation on the Earth?

A. 28 days.
B. 14 days.
C. 4 to 8 minutes.
D. 20 to 40 hours.

We can see a sunspot in the amount of time it takes light to travel from the Sun to the Earth – 8 minutes. But this specific question asks about the slower moving *sunspot charged particles* that lumber toward Earth as part of the solar wind, taking as long as *20 to 40 hours* to begin to disturb radio wave propagation down here on the ham bands. This means we have an almost 2 days "heads up" alert that band conditions may be changing. Tune into WWV at 18 minutes past the hour and listen to the 30-second solar report.

WWV 5 MHz   Best at night   WWV 15 MHz   Best days
WWV 10 MHz   Day and night   WWV & WWVH 20 MHz Some days

You can usually tune in WWV 10 MHz during the days, but only before a big sunspot event finally hits here on Earth 20 to 40 hours later. When disruption occurs, WWV sometimes will fade out completely! **ANSWER D.**

G3A14 How are radio communications usually affected by the charged particles that reach the Earth from solar coronal holes?

A. HF communications are improved.
B. HF communications are disturbed.
C. VHF/UHF ducting is improved.
D. VHF/UHF ducting is disturbed.

A solar coronal hole can be seen as a dark spot on the face of the Sun emitting charged particles into the solar wind. If you took your thermometer up there, there would be an extreme temperature drop within the eruption. You can actually track sunspots as they rotate around the Sun on a 27.5-day cycle. There are usually a pair of sunspots – a pair with a positive magnetic north field, and a pair with a negative south field. Just ask Galileo – he was the first to observe these sunspots in 1610. Sunspots will normally disrupt high-frequency communications, but for the joy of 6 and 2 meter operators, sunspots can create aberrations in the magnetosphere that will cause VHF long-range auroral band openings. But on HF, *coronal holes may lead to poor band conditions*. **ANSWER B.**

G3A07 At what point in the solar cycle does the 20 meter band usually support worldwide propagation during daylight hours?
　　A. At the summer solstice.
　　B. Only at the maximum point of the solar cycle.
　　C. Only at the minimum point of the solar cycle.
　　D. At any point in the solar cycle.

Around 2015, we should reach the peak of solar cycle 24. The higher bands like 6, 10, 12, and 15 meters will be bursting with early morning and late afternoon skywave action. *There is always something going on over on the 20 meter band*. The lower bands, like 40, 60, 75, and 160 meters will continue to give us nighttime skywave excitement. **ANSWER D.**

G3C13 What is Near Vertical Incidence Skywave (NVIS) propagation?
　　A. Propagation near the MUF.
　　B. Short distance HF propagation using high elevation angles.
　　C. Long path HF propagation at sunrise and sunset.
　　D. Double hop propagation near the LUF .

Let's imagine you going out on a camping trip into a valley 200 miles away from your buddy's house. With a normal dipole up about a half wave length, your 40 meter signal is literally skipping over his QTH. A neat way to "pull in" your first skywave hop is to lower your dipole close to the ground just high enough so no one can touch it. You will find that your first skywave hop now arrives at your friend's house loud and clear. Dramatically *lowering an antenna* on purpose close to the Earth will develop a higher elevation angle of radiation, causing your skip distance to "pull in" to establish *shorter-than-normal skywave communications*. **ANSWER B.**

Website Resources

▼ IF YOU'RE LOOKING FOR	▼ THEN VISIT
antennas, DSP, and more	www.amcominc.com
THE place for QSO cards	www.w4mpy.com
DX reference guide	www.ac6v.com
amateur radio satellite operations	www.amsat.org
the latest news about ham radio	www.arnewsline.org
microphones, headsets, and more	www.heilsound.com
digital ham radio accessories and more	www.packetradio.com
Disaster preparedness & emergency comms info	www.fema.gov
NASA research site with cool articles	www.grc.nasa.gov
DSP, antenna analyzers, and more	www.timewave.com
HF modems for digital modes	www.halcomm.com
HF modems and gear for digital modes	www.kantronics.com
digital amateur radio organization	www.tapr.org
azimuthal maps, solar information, and more	www.wm7d.net
software defined radios	www.flexradio.com

TRACK 1

Your HF Transmitter

Elmer Point: *Ham radio single sideband transceivers in the $1800 range will likely offer adjustable digital signal processing filters to perfectly set transmit audio characteristics and bandwidth as well as receive bandwidth selections.* This is why I always recommend buying new HF transceivers that now include new digital network filters.

G8A01 What is the name of the process that changes the envelope of an RF wave to carry information?

A. Phase modulation.

B. Frequency modulation.

C. Spread spectrum modulation.

D. Amplitude modulation.

A type of modulation that changes the *envelope (amplitude)* of an RF wave is called *amplitude modulation (AM).* **ANSWER D.**

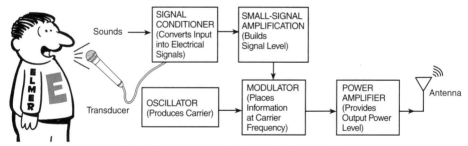

Block Diagram of a Basic Radio Transmitter
Source: *Basic Communications Electronics,* Hudson & Luecke,
© 1999, Master Publishing, Inc., Niles, Illinois

G8A05 What type of modulation varies the instantaneous power level of the RF signal?

A. Frequency shift keying.

B. Pulse position modulation.

C. Frequency modulation.

D. Amplitude modulation.

The instantaneous power level of the signal varies with *amplitude modulation*.
ANSWER D.

G7C02 **Which circuit is used to combine signals from the carrier oscillator and speech amplifier and send the result to the filter in a typical single-sideband phone transmitter?**

A. Discriminator.

B. Detector.

C. IF amplifier.

D. Balanced modulator.

In an SSB transceiver, the *balanced modulator* processes the signal from the carrier oscillator and the speech amplifier and sends it on to the filter. **ANSWER D.**

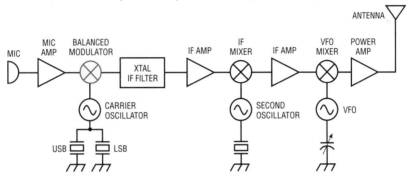

SSB Transmitter Block Diagram

G8A12 **What signal(s) would be found at the output of a properly adjusted balanced modulator?**

A. Both upper and lower sidebands.

B. Either upper or lower sideband, but not both.

C. Both upper and lower sidebands and the carrier.

D. The modulating signal and the unmodulated carrier.

Just as the name implies, the *"balanced modulator"* in your new single-sideband General Class ham transceiver includes *both the upper and lower sidebands* at its output, with the carrier reduced to near zero. Then the signal goes into a sideband filter, and you end up with (normally) lower sideband on 160, 75, and 40 meters, and with upper sideband found on the brand new 60 meter, 5 MHz channels, along with upper sideband on 20, 17, 15, 12, 10, and 6 meters, too. Only brand new equipment offers 60 meter, 5 MHz capabilities, so check with your local radio store to see how you can modify an older SSB to work on any one of the new 5 channels at 5 MHz, double checking that your radio is switched to upper sideband. **ANSWER A.**

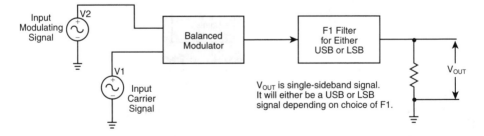

Filtering an SSB Signal

G7C01 Which of the following is used to process signals from the balanced modulator and send them to the mixer in a single-sideband phone transmitter?

A. Carrier oscillator.　　　　　C. IF amplifier.
B. Filter.　　　　　　　　　　　D. RF amplifier.

The *balanced modulator* produces an upper and lower sideband signal, with the carrier balanced out into suppression. The upper and lower sideband signals are then fed *into* a narrow *filter* network which cancels out either the upper or the lower sideband. Modern high frequency ham transceivers now incorporate menu items that allow you to select the shape of the remaining sideband signal, which is then fed to the driver amplifier. The more expensive the ham radio base station, the more elaborate the filter networks. **ANSWER B.**

G8A07 Which of the following phone emissions uses the narrowest frequency bandwidth?

A. Single sideband.　　　　　　C. Phase modulation.
B. Double sideband.　　　　　　D. Frequency modulation.

The modern high frequency ham radio transceiver may include *single sideband* filter networks, which *decrease the amount of spectrum* that the signal occupies. While the resulting signal sounds void of base response, its narrow 2.6 kHz of bandwidth really punches through background noise at the other end of the circuit. If you plan to do a lot of contesting with a huge antenna system, consider a step up to a $1500 high frequency base station with selectable transmit voice bandwidths. **ANSWER A.**

G8A06 What is one advantage of carrier suppression in a single-sideband phone transmission?

A. Audio fidelity is improved.
B. Greater modulation percentage is obtainable with lower distortion.
C. The available transmitter power can be used more effectively.
D. Simpler receiving equipment can be used.

Carrier suppression allows *additional power* to be placed in each sideband. **ANSWER C.**

G4D01 What is the purpose of a speech processor as used in a modern transceiver?

A. Increase the intelligibility of transmitted phone signals during poor conditions.
B. Increase transmitter bass response for more natural sounding SSB signals.
C. Prevent distortion of voice signals.
D. Decrease high-frequency voice output to prevent out of band operation.

The *speech processor* on modern transceivers should be used with caution. Too much processing may cause your signal to go into distortion. But just a little bit of speech processing could increase your average power and *increase the intelligibility* of your transmitted voice signals. When conditions are good, turn the processor OFF! Speech processing is like sending an e-mail in all upper case letters! TURN IT OFF! **ANSWER A.**

G4D02 Which of the following describes how a speech processor affects a transmitted single sideband phone signal?
A. It increases peak power.
B. It increases average power.
C. It reduces harmonic distortion.
D. It reduces intermodulation distortion.

Turning on your transceiver's *speech processor* will not increase PEP output power if you are 100 percent modulated. It will only *increase the average power*, and many times makes your signal sound "too hot" for comfort. Stay off that speech processor button unless it's absolutely necessary. **ANSWER B.**

G8A09 What control is typically adjusted for proper ALC setting on an amateur single sideband transceiver?
A. The RF clipping level.
B. Transmit audio or microphone gain.
C. Antenna inductance or capacitance.
D. Attenuator level.

That new high frequency transceiver you are going to reward yourself with when you pass your exam has a built in transmit safeguard called *Automatic Level Control (ALC)*. This is factory preset, and it acts as an electronic governor to throttle back *transmit power* in case you overdrive input circuits with too much *mic gain* or too much computer drive for data. You can monitor ALC action on the rig's multi-meter. Look for a small, now-and-then jiggle of the ALC indicator, which will indicate proper drive levels. If you overdrive the ALC section of your transmitter, you'll see the ALC meter going up into the red danger area. If this happens, immediately back off mic gain or computer drive power to the audio input section of your transmitter. **ANSWER B.**

G4A05 What is a purpose of using Automatic Level Control (ALC) with a RF power amplifier?
A. To balance the transmitter audio frequency response.
B. To reduce harmonic radiation.
C. To reduce distortion due to excessive drive.
D. To increase overall efficiency.

If you plan to operate digital modes, your computer can take these bizarre sounds and stream messages on the screen. Most computers can take a direct connection between audio out on your transceiver to the microphone input to the computer's sound card. On transmit, you'll want to invest in a relatively inexpensive hardware accessory that lets you fine adjust the input drive to your high frequency radio. Monitor the automatic level control readout on your transceiver to insure that you don't overdrive the modulation input. You should see just a slight wiggle of the indicator showing that *ALC is keeping your computer and interface from overdriving* the transmitter and *the linear amplifier*, which will insure *no distortion* due to excessive drive. The ALC output jack on your new transceiver needs to be tied into the ALC input jack on that linear amplifier. This gives your equipment a "handshake" when the drive level is correctly set. Without this simple plug-in connection between amplifier and rig, the amp could get overdriven by the rig and go into thermal melt down. Be sure to hook up that needed ALC "handshake" cable between rig and amp. **ANSWER C.**

G4A07 What condition can lead to permanent damage when using a solid-state RF power amplifier?
A. Exceeding the Maximum Usable Frequency.
B. Low input SWR.
C. Shorting the input signal to ground.
D. Excessive drive power.
Give your new high frequency radio plenty of ventilation. It has its own built-in fan, but make sure you don't obstruct airflow. What can kill an external power amplifier with solid-state components, and what can kill that brand new HF radio, is going into the digital mode and sending ultra-long messages where the radio gets red hot and ultimately fries a power amplifier output device. *Don't run excessive drive power* when operating in the digital modes! Keep your cool! **ANSWER D.**

G8A08 Which of the following is an effect of over-modulation?
A. Insufficient audio.   C. Frequency drift.
B. Insufficient bandwidth.   D. Excessive bandwidth.
If you *over-modulate* on high frequency, you're going to spread your signal over a *wider bandwidth* than a properly modulated HF signal. If someone says you are "splattering," turn down the mic gain and turn off the speech processor! **ANSWER D.**

G4D03 Which of the following can be the result of an incorrectly adjusted speech processor?
A. Distorted speech.   C. Excessive background pickup.
B. Splatter.   D. All of these choices are correct.
Most newer high frequency ham transceivers have a speech processor button that will either turn processing on or off. TURN THE SPEECH PROCESSOR OFF! *Speech processing* brings in *excessive background pickup*, may cause your *signal to sound distorted*, and may also cause it to *splatter to adjacent frequencies*. Leave your speech processor turned off as a new General Class ham! **ANSWER D.**

G8A10 What is meant by flat-topping of a single-sideband phone transmission?
A. Signal distortion caused by insufficient collector current.
B. The transmitter's automatic level control is properly adjusted.
C. Signal distortion caused by excessive drive.
D. The transmitter's carrier is properly suppressed.
If you turn the microphone gain too high on SSB, that brand new rig will *sound distorted*. The signal waveform shown on an oscilloscope has the top clipped off so it has a flat top. **ANSWER C.**

Elmer's Oscilloscope Waveform
Showing "Flattopping"

G4B15 What type of transmitter performance does a two-tone test analyze?
A. Linearity.
B. Carrier and undesired sideband suppression.
C. Percentage of frequency modulation.
D. Percentage of carrier phase shift.

The faithful reproduction of your voice over single-sideband depends on equipment with excellent linearity. With new digital signal processing techniques on modulation, plus a new family of after-market, high-performance microphones, the *2-tone test* viewing the wave forms on an oscilloscope is a great way to test for *transmitter linearity*. But you know, even though the 2-tone test is your best correct answer for linearity in the transmitter, I always like to ask fellow hams simply how I sound coming over their radio. No oscilloscope needed! **ANSWER A.**

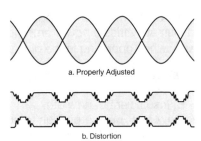

a. Properly Adjusted

Two-Tone Test

b. Distortion

G4B16 What signals are used to conduct a two-tone test?
A. Two audio signals of the same frequency shifted 90-degrees.
B. Two non-harmonically related audio signals.
C. Two swept frequency tones.
D. Two audio frequency range square wave signals of equal amplitude.

More technical hams who regularly service their own equipment have developed 2 non-harmonically related audio tones in a little "warbler" box that may either be plugged into the transmitter mike input, or the mike held up to the little warbler speaker. *The 2 tones must not be harmonically related* in order to provide the best test. Listen to this sound on my audio CD course. **ANSWER B.**

G4B01 What item of test equipment contains horizontal and vertical channel amplifiers?
A. An ohmmeter.
B. A signal generator.
C. An ammeter.
D. An oscilloscope.

An oscilloscope is your best piece of test equipment if you are a technical amateur operator. But it takes skill to work a "scope," so don't buy one unless you know how to use it. The *oscilloscope has horizontal- and vertical-channel amplifiers*. **ANSWER D.**

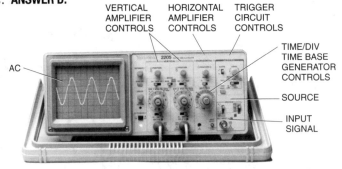

VERTICAL AMPLIFIER CONTROLS   HORIZONTAL AMPLIFIER CONTROLS   TRIGGER CIRCUIT CONTROLS

TIME/DIV TIME BASE GENERATOR CONTROLS

AC

SOURCE

INPUT SIGNAL

A Typical Oscilloscope Showing an AC Waveform
Source: *Basic Electronics* © 1994, Master Publishing, Inc., Niles, Illinois

G4B03 Which of the following is the best instrument to use when checking the keying waveform of a CW transmitter?

A. An oscilloscope.
B. A field-strength meter.
C. A sidetone monitor.
D. A wavemeter.

The oscilloscope is your best instrument to check for transmit signal quality. When magazine editors review the *waveform of a CW signal* or a two-tone test, they usually show photographs of the *oscilloscope* display. **ANSWER A.**

G4B04 What signal source is connected to the vertical input of an oscilloscope when checking the RF envelope pattern of a transmitted signal?

A. The local oscillator of the transmitter.
B. An external RF oscillator.
C. The transmitter balanced mixer output.
D. The attenuated RF output of the transmitter.

We take the *attenuated RF output* of the transmitter and couple it to the vertical input of an oscilloscope to check the quality of the transmitted signal. **ANSWER D.**

G4B02 Which of the following is an advantage of an oscilloscope versus a digital voltmeter?

A. An oscilloscope uses less power.
B. Complex impedances can be easily measured.
C. Input impedance is much lower.
D. Complex waveforms can be measured.

When you whistle into your SSB microphone and look at the resulting wave forms on an oscilloscope, you will see that only an *oscilloscope can truly represent complex voice wave forms* so that they can be precisely measured. **ANSWER D.**

G4A09 Why is a time delay sometimes included in a transmitter keying circuit?

A. To prevent stations from talking over each other.
B. To allow the transmitter power regulators to charge properly.
C. To allow time for transmit-receive changeover operations to complete properly before RF output is allowed.
D. To allow time for a warning signal to be sent to other stations.

Older tube-type high frequency equipment incorporated a transmit/receive relay to toggle output and input antenna connections. Newer HF transceivers have gone solid state with almost instantaneous *switching between receiving and transmitting*. In order to accommodate an external linear amplifier, you may need to review the *time delay transmitter keying circuit* to allow the amplifier to switch on just before it begins to receive transmit power from the transceiver. **ANSWER C.**

G7B13 What is the reason for neutralizing the final amplifier stage of a transmitter?

A. To limit the modulation index.
B. To eliminate self-oscillations.
C. To cut off the final amplifier during standby periods.
D. To keep the carrier on frequency.

After new tubes have been installed in a powerful amplifier, follow the steps in the instruction manual for *neutralization*, which *will eliminate* the possibility of *self-oscillation*. **ANSWER B.**

G8A03 What is the name of the process which changes the frequency of an RF wave to convey information?

A. Frequency convolution.   C. Frequency conversion.
B. Frequency transformation.   D. Frequency modulation.

A type of modulation that changes the frequency of an RF wave is called *frequency modulation (FM)*. **ANSWER D.**

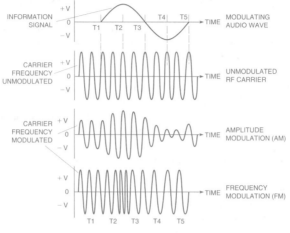

Modulation of RF Carrier

G8A11 What happens to the RF carrier signal when a modulating audio signal is applied to an FM transmitter?

A. The carrier frequency changes proportionally to the instantaneous amplitude of the modulating signal.
B. The carrier frequency changes proportionally to the amplitude and frequency of the modulating signal.
C. The carrier amplitude changes proportionally to the instantaneous frequency of the modulating signal.
D. The carrier phase changes proportionally to the instantaneous amplitude of the modulating signal.

When you get your new General Class license, you still will be working with the gang on 2 meters and 440 MHz FM. The way your FM transceiver changes your voice to a signal on the air is the carrier frequency changes proportionately to the instantaneous amplitude of the modulating signal. Now say 5 times, out loud, "Frequency changes instantaneous amplitude," and then again, "Frequency changes instantaneous amplitude," and then again, *"Frequency changes instantaneous amplitude."* Got it? **ANSWER A.**

G8B04 What is the name of the stage in a VHF FM transmitter that generates a harmonic of a lower frequency signal to reach the desired operating frequency?

A. Mixer.   C. Pre-emphasis network.
B. Reactance modulator.   D. Multiplier.

Inside a VHF FM transmitter is a low-power oscillator that operates at HF levels. It's the job of the *multiplier* to select a harmonic of this signal to produce the desired operating frequency. **ANSWER D.**

G8B06 What is the total bandwidth of an FM-phone transmission having a 5 kHz deviation and a 3 kHz modulating frequency?

A. 3 kHz.

B. 5 kHz.

C. 8 kHz.

D. 16 kHz.

You can calculate this answer by multiplying 2 times the sum of the deviation and highest audio modulating frequency. The deviation is 5 kHz plus 3 kHz of audio, which gives a sum of 8 kHz. Two times 8 kHz equals *16 kHz*. This would be the *total bandwidth* of the FM phone transmission. **ANSWER D.**

G8B07 What is the frequency deviation for a 12.21-MHz reactance-modulated oscillator in a 5-kHz deviation, 146.52-MHz FM-phone transmitter?

A. 101.75 Hz.

B. 416.7 Hz.

C. 5 kHz.

D. 60 kHz.

This is an easy ratio problem. First, let's determine the frequency multiplication factor of the transmitter. Divide the 12.21 oscillator frequency into the output at 146.52. This gives us a multiplication factor of 12. If there is 5 kHz deviation at the transmitter, 1/12th deviation at the oscillator input is 12 divided into 5000 Hz, with an answer of 416.66 Hz, rounded to *416.7 Hz*. **ANSWER B.**

G8A04 What emission is produced by a reactance modulator connected to an RF power amplifier?

A. Multiplex modulation.

B. Phase modulation.

C. Amplitude modulation.

D. Pulse modulation.

A *reactance modulato*r produces *phase modulation*, which is used for both phase modulation and frequency modulation. **ANSWER B.**

G8A02 What is the name of the process that changes the phase angle of an RF wave to convey information?

A. Phase convolution.

B. Phase modulation.

C. Angle convolution.

D. Radian inversion.

A type of modulation that changes the phase of an RF wave is called *phase modulation (PM)*. **ANSWER B.**

G7B08 How is the efficiency of an RF power amplifier determined?

A. Divide the DC input power by the DC output power.

B. Divide the RF output power by the DC input power.

C. Multiply the RF input power by the reciprocal of the RF output power.

D. Add the RF input power to the DC output power.

You have your brand new high frequency transceiver hooked up to a watt meter and a dummy load. The sales brochure indicates that new rig offers 250 watts input. Yet, when you key the microphone and look at the watt meter into the dummy load and say the word "FOOOUUUURRRRR," the needle barely goes to 100 watts output. Hey, what's going on here? Power input to the final transistor block yields about 50% efficiency, or slightly less, using voice. Some of the power goes up as heat, a little bit of power may drive other stages, and the resulting output power will always be much less than input power. *To calculate the efficiency of your new high frequency transceiver, you divide the RF output power by the DC input power*. The RF output power is easily seen on the watt meter, and if you can whistle in the mike and get 100 watts out, your rig is working as it should. If you want to get scientific and calculate input power, you will need to calculate voltage times transistor block current. But just remember the

---

formula: efficiency = power output divided by power input. And just remember, high frequency output power as measured with an inline watt meter for voice will dance around 60 watts, and this is plenty of power to work the world with your new General Class privileges. **ANSWER B.**

Linear RF power amplifier.

Elmer Point: *Don't run out and buy a linear amplifier. The 100-watt output from your HF ham transceiver is plenty powerful enough to work the world. If you want to blast your signal out stronger, go for a directional antenna. The directional antenna also will increase incoming received signal strength – something that linear amplifier can't do. Linear amps are for those hams who have been on the air for at least a year, and who put up a 60 foot tower with a big directional beam on top. What the amplifier will do is increase their ability to bust through a pileup. Pileups occur when numerous hams are simultaneously trying to get their call signs through to a rare DX station during a contest. When you're just getting started on the General Class airwaves, steer clear of contest operation where pileups take place.*
☞ Visit: www.ameritron.com

G5B06 What is the output PEP from a transmitter if an oscilloscope measures 200 volts peak-to-peak across a 50-ohm dummy load connected to the transmitter output?
A. 1.4 watts.   C. 353.5 watts.
B. 100 watts.   D. 400 watts.

Here is an Ohm's Law circle problem, and notice we have added power in addition to resistance, voltage, and current to the Ohm's Law circle shown on page 120. You will need to memorize some of the equations you pull from the circle, and you have 12 formulas and equations to start your search for the right one to get your calculation job done. For this question, the formula that applies is $P = E^2 \div R$.

This particular question has you looking at an oscilloscope that indicates 200 volts peak-to-peak across a 50-ohm dummy load that is tied in to your Granddad's old radio. Let's first reduce the 200 volts peak-to-peak down to 100 volts, just peak. Next, our calculations are for peak envelope power, which is the AVERAGE of the output power, as defined by the FCC. To calculate peak to average, multiply peak envelope voltage by 0.707. Now apply the formula: power equals the square of the voltage divided by the load resistance. Here are the calculator keystrokes: $100 \times 0.707 = 70.7 \times 70.7 = 4998.49 \div 50 = 99.7$, rounded to *100 watts*. Now that wasn't too tough, was it? **ANSWER B.**

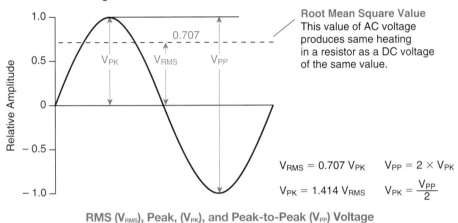

Root Mean Square Value
This value of AC voltage produces same heating in a resistor as a DC voltage of the same value.

$$V_{RMS} = 0.707 \; V_{PK} \qquad V_{PP} = 2 \times V_{PK}$$
$$V_{PK} = 1.414 \; V_{RMS} \qquad V_{PK} = \frac{V_{PP}}{2}$$

RMS (V_{RMS}), Peak, (V_{PK}), and Peak-to-Peak (V_{PP}) Voltage

---

G5B14 What is the output PEP from a transmitter if an oscilloscope measures 500 volts peak-to-peak across a 50-ohm resistor connected to the transmitter output?

A. 8.75 watts.   C. 2500 watts.
B. 625 watts.   D. 5000 watts.

Again, take half of the peak-to-peak voltage to obtain peak voltage. Multiply 250×0.707, square the result (176.75) by multiplying it by itself, and divide by 50. Here are the calculator keystrokes: Clear, Clear $250 \times 0.707 = 176.75 \times 176.75 = 31,240.56 \div 50 = 624.81$, which rounds off to *625 watts*. **ANSWER B.**

G5B11 What is the ratio of peak envelope power to average power for an unmodulated carrier?

A. 0.707.   C. 1.414.
B. 1.00.   D. 2.00.

An unmodulated carrier is simultaneously peak and average. So the ratio of peak envelope power to average power on a steady unmodulated carrier is simply *1*. **ANSWER B.**

G5B13 What is the output PEP of an unmodulated carrier if an average reading wattmeter connected to the transmitter output indicates 1060 watts?

A. 530 watts.   C. 1500 watts.
B. 1060 watts.   D. 2120 watts.

Remember, a steady carrier illustrates both peak power as well as average power. If your watt meter reads *1060 watts* average, this is the same power as peak envelope power. **ANSWER B.**

G5B01 A two-times increase or decrease in power results in a change of how many dB?
A. Approximately 2 dB.   C. Approximately 6 dB.
B. Approximately 3 dB.   D. Approximately 12 dB.
The decibel is used to describe a change in power levels. It is a measure of the ratio of power output to power input. A *two-times increase* results in a change of *3 dB*. **ANSWER B.**

If $dB = 10 \log_{10} \dfrac{P_1}{P_2}$

then what power ratio is 20 dB?

$20 = 10 \log_{10} \dfrac{P_1}{P_2}$

$\dfrac{20}{10} = \log_{10} \dfrac{P_1}{P_2}$

$2 = \log_{10} \dfrac{P_1}{P_2}$

Remember: logarithm of a number is the exponent to which the base must be raised to get the number.

$\therefore 10^2 = \dfrac{P_1}{P_2}$

$100 = \dfrac{P_1}{P_2}$

Or $P_1 = 100\ P_2$

20 dB means P_1 is 100 times P_2

dB	$\dfrac{P_1}{P_2}$
3	2
6	4
10	10
20	100
30	1000
40	10000
50	10^5
60	10^6

Definition of a Decibel
Source: *The Technology Dictionary*,© 1987 Master Publishing, Inc., Niles, IL

G4D07 How much must the power output of a transmitter be raised to change the S meter reading on a distant receiver from S8 to S9?
A. Approximately 1.5 times.   C. Approximately 4 times.
B. Approximately 2 times.   D. Approximately 8 times.
Seeing an S meter change from S8 to S9 is an increase of a single S unit. One S unit is 6 dB, and 6 dB is a *4-times change*. **ANSWER C.**
Here is how the dB system for power works:
0 dB = 0 times change
3 dB = 2 times change
6 dB = 4 times change
9 dB = 8 times change
10 dB = 10 times change

G4D05 How does an S meter reading of 20 dB over S-9 compare to an S-9 signal, assuming a properly calibrated S meter?
A. It is 10 times weaker.   C. It is 20 times stronger.
B. It is 20 times weaker.   D. It is 100 times stronger.
An S meter reading of *20 db over S-9*, compared to just an S-9 signal, illustrates the signal is *100 times stronger*. **ANSWER D.**

G7B11 For which of the following modes is a Class C power stage appropriate for amplifying a modulated signal?
A. SSB.   C. AM.
B. CW.   D. All of these choices are correct.
The Class C amplifier is ideal for non-linear amplification of Morse code (CW) plus many digital emissions. Varying amplitude emissions like SSB or AM would not sound great if the amplifier is operating in the *Class C* region. But for *CW*, it will work great. **ANSWER B.**

G7B12 Which of these classes of amplifiers has the highest efficiency?

A. Class A.
B. Class B.
C. Class AB.
D. Class C.

Yes, Grandpa, I will learn the code! CW (Morse code) will always continue to be a popular mode for ham operators, even though the code test has been completely eliminated. A small CW-only transceiver operating Class C offers high efficiency with little current being consumed from the battery in between dots and dashes. *Class C, high efficiency*. Class A, low distortion. **ANSWER D.**

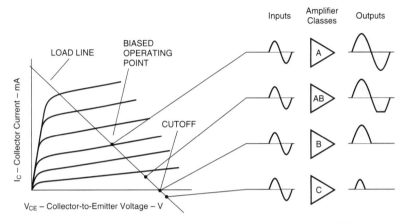

Various Classes Various Classes of Transistorized Amplifiers

G7B10 Which of the following is a characteristic of a Class A amplifier?

A. Low standby power.
B. High Efficiency.
C. No need for bias.
D. Low distortion.

The *Class A amplifier* in your new high frequency transceiver is many times found within the microphone input stage, where *low distortion* is an absolute requirement. Class A amplifiers are continuously drawing a small amount of current to provide the utmost linear operation. **ANSWER D.**

G7B14 Which of the following describes a linear amplifier?

A. Any RF power amplifier used in conjunction with an amateur transceiver.
B. An amplifier in which the output preserves the input waveform.
C. A Class C high efficiency amplifier.
D. An amplifier used as a frequency multiplier.

Don't run right out and buy a $1000 linear amplifier to boost the power output on your $700-$900 high frequency transceiver. Linear amplifiers will indeed pump up your *output waveform exactly as it appears to the amplifier input*. It is unwise to run any linear amplifier on high frequency mobile or on a simple home antenna until you see that 100 powerful watts can work all around the world.
ANSWER B.

This linear power amp can boost the 1- to 5-watt output of a hand-held up to 30 watts.

G4A08 What is the correct adjustment for the load or coupling control of a vacuum tube RF power amplifier?
A. Minimum SWR on the antenna.
B. Minimum plate current without exceeding maximum allowable grid current.
C. Highest plate voltage while minimizing grid current.
D. Maximum power output without exceeding maximum allowable plate current.

On the load control on that old power amp, seasoned hams will tell you to "tune for maximum smoke." Well, not really – but tune for a *maximum power output* reading, making sure you do not exceed 1500 watts output. Most older amps will do 1,000 watts and not much more. **ANSWER D.**

G4A04 What reading on the plate current meter of a vacuum tube RF power amplifier indicates correct adjustment of the plate tuning control?
A. A pronounced peak.
B. A pronounced dip.
C. No change will be observed.
D. A slow, rhythmic oscillation.

Vacuum tube linear amplifiers are always a welcome ham radio "accessory" for the shack – especially if your Granddad gives you his old linear amplifier! When hooked up to a properly-tuned antenna, the plate tuning control will usually show a pronounced dip on the amplifier's big meter. If you don't see a pronounced dip, time to see what went wrong with your antenna system. The more resonant your tuned antenna is, the more pronounced the *dip in your plate current meter reading*. **ANSWER B.** ☞ **Visit: www.radioblvd.com**
www.heathkit-museum.com

Your Receiver

G7C07 What is the simplest combination of stages that implement a superheterodyne receiver?

A. RF amplifier, detector, audio amplifier.

B. RF amplifier, mixer, discriminator.

C. HF oscillator, mixer, detector.

D. HF oscillator, pre-scaler, audio amplifier.

The early (before you were born) ham radio high frequency receiver operated as TRF – tuned radio frequency. The receiver was nice and sensitive, but not all that stable. Today, our modern ham radio receiver uses superheterodyne technology with intermediate frequency amplifiers, which include deep, narrow-frequency filters offering plenty of gain, excellent frequency stability, and menu-optional selectivity and sensitivity. The term "superheterodyne" indicates a receiver with local oscillators and mixers which recover an incoming signal clean of other signals just above and below the desired signal. New ham transceivers now allow the operator to customize the intermediate frequency band pass filters for just the right amount of bandwidth to clearly hear the incoming weak signal. The simplest combination of stages to describe a *superheterodyne receiver* are the *HF oscillator, the mixer, and the detector*. **ANSWER C.**

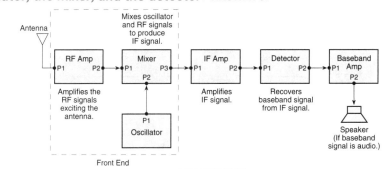

a. Generic Receiver

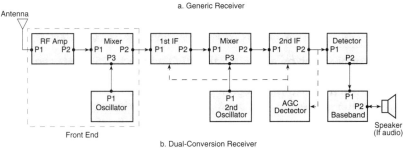

b. Dual-Conversion Receiver

Block Diagrams of Generic and Dual-Conversion AM Receivers
Source: *Basic Communications Electronics,* © 1999 Master Publishing, Inc., Niles, IL

G7C03 What circuit is used to process signals from the RF amplifier and local oscillator and send the result to the IF filter in a superheterodyne receiver?
A. Balanced modulator.   C. Mixer.
B. IF amplifier.   D. Detector.
In an SSB receiver section, it is the *mixer* that processes signals from the RF amplifier and the local oscillator. **ANSWER C.**

G7C04 What circuit is used to combine signals from the IF amplifier and BFO and send the result to the AF amplifier in a single-sideband receiver?
A. RF oscillator.   C. Balanced modulator.
B. IF filter.   D. Product detector.
In an SSB receiver, it is the *product detector* that processes the signal from the IF amplifier and the beat frequency oscillator. The signal then goes on to the audio frequency amplifier. **ANSWER D.**

G8B03 What is another term for the mixing of two RF signals?
A. Heterodyning.   C. Cancellation.
B. Synthesizing.   D. Phase inverting.
When a ham is tuning around the worldwide bands and listens in on a juicy conversation, and then another signal comes in right on top of it, they sometimes refer to this as a "heterodyne." This was a major problem in the early days of AM CB radio. But believe it or not, the inside of your equipment uses *heterodyning* for the positive purpose of *mixing 2 RF signals* near the IF stage. **ANSWER A.**

G8B01 What receiver stage combines a 14.250 MHz input signal with a 13.795 MHz oscillator signal to produce a 455 kHz intermediate frequency (IF) signal?
A. Mixer.   C. VFO.
B. BFO.   D. Discriminator.
When you see that word "combines," think of the *mixer* section of a receiver stage. **ANSWER A.**

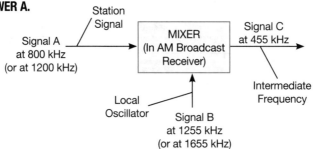

Block diagram of an AM broadcast receiver mixer.
Source: *Basic Communications Electronics*, © 1999 Master Publishing, Inc., Niles, IL

G8B02 If a receiver mixes a 13.800 MHz VFO with a 14.255 MHz received signal to produce a 455 kHz intermediate frequency (IF) signal, what type of interference will a 13.345 MHz signal produce in the receiver?
A. Quadrature noise.   C. Mixer interference.
B. Image response.   D. Intermediate interference.
In strong signal areas where there may be local transmissions coming in from shortwave stations outside of normal ham band limits, an interference called *"image response"* may develop *at the sum and difference of your intermediate frequency (IF) signal*. 13.800 MHz minus 455 kHz is 13.345 MHz. **ANSWER B.**

G4C12 Which of the following is an advantage of a receiver Digital Signal Processor IF filter as compared to an analog filter?
A. A wide range of filter bandwidths and shapes can be created.
B. Fewer digital components are required.
C. Mixing products are greatly reduced.
D. The DSP filter is much more effective at VHF frequencies.

The new, modern high frequency transceiver usually includes digital signal processing. This allows the user to *create multiple filter bandwidth settings*, along with bandwidth shape for "soft" or "hard" response. This wide range of filter combinations – all digital – really helps subtract background noise. **ANSWER A.**

G7C09 Which of the following is needed for a Digital Signal Processor IF filter?
A. An analog to digital converter.   C. A digital processor chip.
B. A digital to analog converter.   D. All of the these choices are correct.

Inside a DSP integrated circuit is a digital processor that converts analog to digital, a noise subtraction stage, and then digital to analog converter. So, *all of these answers are correct* in the modern DSP IF filter. **ANSWER D.**

G4C11 Which of the following is one use for a Digital Signal Processor in an amateur station?
A. To provide adequate grounding.
B. To remove noise from received signals.
C. To increase antenna gain.
D. To increase antenna bandwidth.

Digital signal processing (DSP) is found in almost all new high frequency amateur radio transceivers. Older HF sets don't have DSP, and this is why I always recommend buying a new HF transceiver. New HF equipment is available for under $600, 100 watts out, 100 channels of memory, DSP included! *DSP* is a great way to *remove noise from received weak signals*. **ANSWER B.**
☞ Visit: www.bhi-ltd.co.uk
www.westmountainradio.com

A DSP Speaker.

Elmer Point: *If you start out with a used, older, HF radio, you can buy an audio DSP speaker system for about $225. You'll be amazed at how the DSP noise subtraction circuit improves the sounds coming from the radio. But better yet, buy a new HF transceiver with DSP filtering in the IF and now you'll really hear a big difference of cleaned-up reception. DSP is not magic, however. The best way to improve reception is to search out whatever it is around your house that is generating all the noise and get it shut down when you begin to operate with those distant, rare, weak-signal stations. Fans and florescent lights are huge noise generators.*

G7C10 How is Digital Signal Processor filtering accomplished?
   A. By using direct signal phasing.
   B. By converting the signal from analog to digital and using digital processing.
   C. By differential spurious phasing.
   D. By converting the signal from digital to analog and taking the difference of mixing products.

The input of a *DSP* circuit *takes an analog signal, converts it to a digital signal* where the noise subtraction takes place, *and restores it to an analog signal* so we can hear it again! **ANSWER B.**

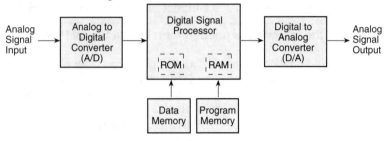

Block Diagram of a Basic Digital Signal Processing (DSP) System
Source: *Basic Communications Electronics,* © 1999 Master Publishing, Inc., Niles, IL

G4A13 What is one reason to use the attenuator function that is present on many HF transceivers?
   A. To reduce signal overload due to strong incoming signals.
   B. To reduce the transmitter power when driving a linear amplifier.
   C. To reduce power consumption when operating from batteries.
   D. To slow down received CW signals for better copy.

On the front of your radio is a button marked "ATT" – attenuator. When you push it in, you'll notice that the reception drops in signal strength. Sometimes there may be several levels of attenuation. Why in the world would you want to lessen reception? Easy answer – at full sensitivity you sometimes may pull in interference from other stations on either side of the station you are listening to, or cause an extremely strong signal to swamp your receiver, called overload. *The attenuator will minimize overload*. Another good trick is to also turn OFF the preamplifier on receive when conditions get crowded on the band. **ANSWER A.**

G4A01 What is the purpose of the "notch filter" found on many HF transceivers?
   A. To restrict the transmitter voice bandwidth.
   B. To reduce interference from carriers in the receiver passband.
   C. To eliminate receiver interference from impulse noise sources.
   D. To enhance the reception of a specific frequency on a crowded band.

Your 40 meter privileges are shared with foreign broadcast stations. From early evening to a couple of hours after Sun-up, you will hear some mighty interesting full-carrier, double-sideband AM shortwave stations coming from the Far East. We can sometimes *dodge the interference*, thanks to your high frequency transceiver's *"notch filter."* The notch filter may be variable to reject a specific frequency "whistle" coming from these far away shortwave stations. Adjust it carefully and the whistle may disappear. Newer transceivers may incorporate automatic notch filters, based on digital signal processing, capable of notching-out several interfering heterodyne signals. **ANSWER B.**

G4C13 Which of the following can perform automatic notching of interfering carriers?
A. Band-pass tuning.
B. A Digital Signal Processor (DSP) filter.
C. Balanced mixing.
D. A noise limiter.
Always buy new ham gear to get started on your new HF privileges. New gear incorporates a circuit called "ANF" on the front panel – automatic notch filter. This is handy on 40 meters where there is a lot of foreign broadcast carrier noise that will drive you crazy without the automatic notch filter. *"ANF" uses digital signal processing* technology to magically eliminate the noise. **ANSWER B.**

G7C08 What type of circuit is used in many FM receivers to convert signals coming from the IF amplifier to audio?
A. Product detector.   C. Mixer.
B. Phase inverter.   D. Discriminator.
In a frequency modulation receiver, signals coming from the intermediate frequency amplifier are fed into the *discriminator*, which acts as a frequency to voltage conversion stage, sending the resulting "decoded" signal on to the audio amplifier. **ANSWER D.**

G4D06 Where is an S meter found?
A. In a receiver.   C. In a transmitter.
B. In an SWR bridge.   D. In a conductance bridge.
The *S meter* on your new HF transceiver ties directly in to the *receiver* section of the radio. Although the large meter movement may illustrate other radio parameters, when it is in the "S meter" setting it is tied in to the receiver's automatic gain control (AGC) signal strength circuitry. On most HF transceivers, there also is a front panel setting for receiver AGC. Set it at "automatic" or "slow" when copying single sideband voice emissions. Set it to "fast" when copying CW. In the "slow" or "automatic" settings, the S meter will faithfully read out incoming signal levels. If the AGC is turned "OFF", the S meter won't budge! Leave it on "slow" or "automatic". **ANSWER A.**

S meter reading on radio.

G4D04 What does an S meter measure?
A. Conductance.   C. Received signal strength.
B. Impedance.   D. Transmitter power output.
High frequency ham transceivers all incorporate an S meter. *S meters measure received signal strength*. You also might be able to use the S meter to determine if your antenna system is working properly. Try this: on 40 meters, you should see an approximately S-2 indication of normal background noise. If you see a higher reading of background noise, this is an indication that your antenna is performing properly. However, if your S meter doesn't budge on 40 meter noise, something is probably wrong with your antenna system. **ANSWER C.**

Website Resources

▼ IF YOU'RE LOOKING FOR	▼ THEN VISIT
Ham Equipment Reviews	www.eham.net
HRO – Ham Radio Outlet	www.hamradio.com
AES – Amateur Radio Supply	www.aesham.com
Largest Ham Accessory Catalog	www.mfjenterprises.com
Advanced Specialties	www.advancedspecialties.net
Alltronics	www.alltronics.com
Amateur Accessories	www.amateuraccessories.com
Asscociated Radio	www.associatedradio.com
Austin Amateur Radio	www.aaradio.com
B&H Sales	www.hamradiocenter.com
Cedar City Sales	www.cedarcitysales.com
Central Utah Electronics Supply	www.electronicspro.com
Communications Products	www.commproducts.net
DBJ Radio & Electronics	www.dbjre.com
GigaParts, Inc.	www.gigaparts.com
HamStop.com	www.hamstop.com
Houston Amateur Radio Supply	www.texasparadise.com/hars
Jun's Electronics	www.hamcity.com
K1CRA Radio Webstore	www.k1cra.com
K-Comm, Inc. – The Ham Store	www.kcomm.biz
KJI Electronics, Inc.	www.kjielectronics.com
Lentini Communications, Inc.	www.lentinicomm.com
R & L Electronics	www.randl.com
Rad-Comm Radio	www.radcomm.bizland.com/rad-comm
Radio City	www.radioinc.com
The Ham Station	www.hamstation.com
Universal Radio Inc	www.universal-radio.com
Waypoint Crusing Solutions	www.waypoints.com
WB0W, Inc.	www.wb0w.com

Note: This list includes many ham radio dealers where you can purchase your HF radio and accessories

Oscillators & Components

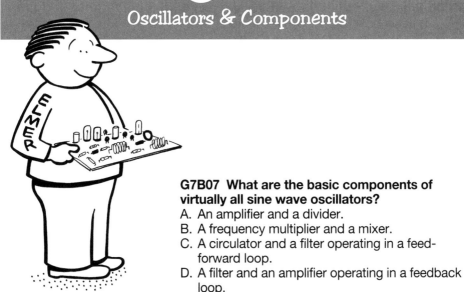

G7B07 What are the basic components of virtually all sine wave oscillators?
A. An amplifier and a divider.
B. A frequency multiplier and a mixer.
C. A circulator and a filter operating in a feed-forward loop.
D. A filter and an amplifier operating in a feedback loop.

To keep an oscillator "in motion," a tuned circuit at one specific frequency will provide the necessary *feedback loop*. Just like keeping Grandpa swinging back and forth on his brand new senior citizen swing set, a small amount of feedback continuously keeps him in motion. **ANSWER D.**

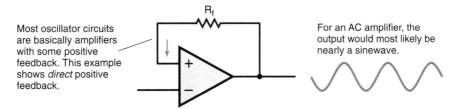

Most oscillator circuits are basically amplifiers with some positive feedback. This example shows *direct* positive feedback.

R_f

For an AC amplifier, the output would most likely be nearly a sinewave.

a. Amplifier with Positive Feedback – Oscillator

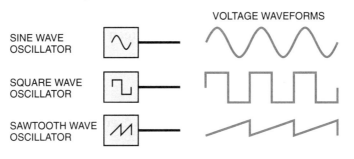

VOLTAGE WAVEFORMS

SINE WAVE OSCILLATOR

SQUARE WAVE OSCILLATOR

SAWTOOTH WAVE OSCILLATOR

b. Oscillator Waveforms

An oscillator is basically an amplifier with positive feedback from output to input.
Source: *Basic Electronics* © 1994, Master Publishing, Inc., Niles, Illinois

G7B09 What determines the frequency of an LC oscillator?
  A. The number of stages in the counter.
  B. The number of stages in the divider.
  C. The inductance and capacitance in the tank circuit.
  D. The time delay of the lag circuit.

An LC oscillator is found in the radio frequency circuits of your new HF transceiver, and consists of a tuned *inductance and capacitance* resonant circuit, found in the tank circuit of that new "Gee Whiz" high frequency transceiver. **ANSWER C.**

G6B10 Which element of a triode vacuum tube is used to regulate the flow of electrons between cathode and plate?
  A. Control grid.   C. Screen grid.
  B. Heater.   D. Trigger electrode.

In the triode vacuum tube, the *control grid* acts like a variable valve to regulate the flow of electrons between the cathode and plate. **ANSWER A.**

High-power vacuum tubes require neutralization.

G6B12 What is the primary purpose of a screen grid in a vacuum tube?
  A. To reduce grid-to-plate capacitance.
  B. To increase efficiency.
  C. To increase the control grid resistance.
  D. To decrease plate resistance.

Yes, hams still run rigs with vacuum tubes. A diode tube has an anode and a cathode that may incorporate direct or indirect heating. The diode tube is normally used in power supplies. The triode tube incorporates a control grid and, like a variable amplifier, the control grid acts as a valve. Next is the tetrode tube, and it adds a *screen grid* that *reduces grid-to-plate capacitance* and collects secondary emissions from the plate, which will slightly increase screen grid current. **ANSWER A.**

G6B11 Which of the following solid state devices is most like a vacuum tube in its general operating characteristics?

A. A bipolar transistor.
B. A Field Effect Transistor.
C. A tunnel diode.
D. A varistor.

The *Field Effect Transistor (FET)* has high input impedance, and has a gate, drain, and source. The gate is similar to the control grid of a tube, with the gate generating an electric field that increases or decreases the amount of current flow. **ANSWER B.**

G6B08 Why must the cases of some large power transistors be insulated from ground?

A. To increase the beta of the transistor.
B. To improve the power dissipation capability.
C. To reduce stray capacitance.
D. To avoid shorting the collector or drain voltage to ground.

Be careful where you place the power supply that changes household power to 12 volts DC. Older transformer-type power supplies mount their power transistors on the large heat sinks that are exposed to the open air. If the supply's power transistor is accidentally pushed up against something else that is *ground*, you *will short out the collector voltage* and likely destroy the power transistor. **ANSWER D.**

G6B07 What are the stable operating points for a bipolar transistor used as a switch in a logic circuit?

A. Its saturation and cut-off regions.
B. Its active region (between the cut-off and saturation regions).
C. Its peak and valley current points.
D. Its enhancement and deletion modes.

When the bipolar transistor reaches saturation, collector and emitter base junctions are forward-biased. When the bipolar transistor is used as a switch in a logic circuit, an extremely small change in collector-base voltage will cause a large change in collector current, allowing the transistor to switch between cut-off and saturation within collector current specifications to keep the transistor from self-destructing. It is in this *saturation and cut-off region* that the transistor will act as a switch. **ANSWER A.**

G6B09 Which of the following describes the construction of a MOSFET?

A. The gate is formed by a back-biased junction.
B. The gate is separated from the channel with a thin insulating layer.
C. The source is separated from the drain by a thin insulating layer.
D. The source is formed by depositing metal on silicon.

Many times, we find the MOSFET transistor in the "front end" of your new modern HF transceiver. The MOSFET is a Metal-Oxide Semiconductor Field-Effect Transistor in which the *gate is separated from the channel with an extremely thin insulating layer*. Manufacturers usually install a gate-protective Zener diode, which prevents the gate insulation from being punctured by small static charges or excessive voltages from an event like a nearby lightning strike. **ANSWER B.**

G6B03 What is the approximate junction threshold voltage of a germanium diode?

A. 0.1 volt.   C. 0.7 volts.
B. 0.3 volts.   D. 1.0 volts.

The *Germanium diode* is sometimes found in the detector stage of a receiver. Their low forward voltage drop is minimal, and the approximate junction *threshold voltage* of this diode is *0.3 volts*. These signal diodes are sensitive and can burn out if you accidentally overheat them while soldering in a replacement. **ANSWER B.**

G6B05 What is the approximate junction threshold voltage of a conventional silicon diode?

A. 0.1 volt.   C. 0.7 volts.
B. 0.3 volts.   D. 1.0 volts.

We find *silicone diodes* in the rectifier section of your new transceiver's power supply. The silicone diode offers stable operation at high temperatures, a *junction threshold voltage* of about *0.7 volts*, and high reverse resistance along with long-term reliability. **ANSWER C.**

G6B06 Which of the following is an advantage of using a Schottky diode in an RF switching circuit as compared to a standard silicon diode?

A. Lower capacitance.   C. Longer switching times.
B. Lower inductance.   D. Higher breakdown voltage.

The Schottky diode is a fast-switching, point-contact diode with lower capacitance offering faster switching capability. This is an important consideration in HF transceivers to handle many digital modes where fast switching time between transmit and receive is required. The *lower* the *capacitance*, the greater is the advantage of the *Schottky* over a common silicone diode. **ANSWER A.**

G6C07 What is one disadvantage of an incandescent indicator compared to an LED?

A. Low power consumption.   C. Long life.
B. High speed.   D. High power consumption.

Long live the light emitting diode. These are favorites in ham radio transceivers. Luckily, they don't burn out like older incandescent lamps. The older *incandescent* indicator *lamps* also *consume high power levels*. Read carefully – this question asks for the DISadvantage of the incandescent light. **ANSWER D.**

G6C08 How is an LED biased when emitting light?

A. Beyond cutoff.   C. Reverse Biased.
B. At the Zener voltage.   D. Forward Biased.

The *light emitting diode* is always forward biased when it energizes. When current is passed through the PN junction, *forward biased*, the light emitting diode almost instantly turns on. **ANSWER D.**

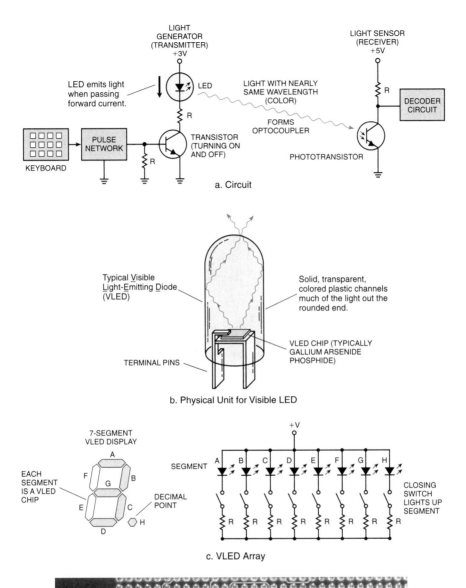

LIGHT
GENERATOR
(TRANSMITTER)
+3V

LED emits light
when passing
forward current.

LED

LIGHT WITH NEARLY
SAME WAVELENGTH
(COLOR)

R

KEYBOARD

PULSE
NETWORK

R

TRANSISTOR
(TURNING ON
AND OFF)

FORMS
OPTOCOUPLER

LIGHT SENSOR
(RECEIVER)
+5V

R

DECODER
CIRCUIT

PHOTOTRANSISTOR

a. Circuit

Typical Visible
Light-Emitting Diode
(VLED)

Solid, transparent,
colored plastic channels
much of the light out the
rounded end.

VLED CHIP (TYPICALLY
GALLIUM ARSENIDE
PHOSPHIDE)

TERMINAL PINS

b. Physical Unit for Visible LED

7-SEGMENT
VLED DISPLAY

A

F   B
    G
E   C

D

EACH
SEGMENT
IS A VLED
CHIP

DECIMAL
POINT

H

+V

SEGMENT   A   B   C   D   E   F   G   H

CLOSING
SWITCH
LIGHTS UP
SEGMENT

R   R   R   R   R   R   R   R

c. VLED Array

An array of LEDs and resistors mounted on a printed circuit board.

G6C09 Which of the following is a characteristic of a liquid crystal display?
A. It requires ambient or back lighting.
B. It offers a wide dynamic range.
C. It has a wide viewing angle.
D. All of these choices are correct.

Many of your new high frequency transceivers use a *liquid crystal display (LCD)*. They require *backlighting* at night, and the monochrome (black and white) liquid crystal displays achieve their own "brilliance" out in the sunshine. Color liquid crystal displays have dramatically improved in direct sunlight visibility and, at night using backlighting, the color LED really looks terrific! **ANSWER A.**

An LCD.

Elmer Point: *If you plan to operate your radio in the field, make sure it offers a liquid crystal display. An amber background with black numbers is best for reading the display in bright sunlight. Color LCD displays are great at night, but they may require some shade during the day. Older HF transceivers may use LED or vacuum florescent display, and these are next to impossible to see when operating Field Day in a tent out in the sunshine. Go for amber LCD with black numbers for best viewing.*

G6C11 What is a microprocessor?
A. A low power analog signal processor used as a microwave detector.
B. A computer on a single integrated circuit.
C. A microwave detector, amplifier, and local oscillator on a single integrated circuit.
D. A low voltage amplifier used in a microwave transmitter modulator stage.

Thanks to microprocessors, your new General Class high frequency transceiver works much like a miniature *computer* with everything *on a single integrated circuit chip*. But just like your home computers, make sure your equipment has enough "breathing room." **ANSWER B.**

G6C02 What is meant by the term MMIC?
A. Multi Megabyte Integrated Circuit.
B. Monolithic Microwave Integrated Circuit.
C. Military-specification Manufactured Integrated Circuit.
D. Mode Modulated Integrated Circuit.

The *monolithic microwave integrated circuit (MMIC)* is a favorite among microwave operators, up on 10 GHz. Up on "X" band, we use MMICs within our microwave equipment as fixed-gain amplifiers with excellent signal to noise ratios. **ANSWER B.**

RF GROUND

RF OUTPUT AND $+V_{cc}$

RF INPUT (DIAGONAL CUT)

Monolithic Microwave Integrated Circuit
Photo Courtesy of Hewlett Packard Co.

G6C06 Which of the following describes an integrated circuit operational amplifier?

A. Digital.

B. MMIC.

C. Programmable Logic.

D. Analog.

We nickname the operational amplifier an *"op amp,"* and this is an *analog integrated circuit device*. The op amp offers high gain over a large range of input frequencies, and is a direct-coupled differential amplifier whose characteristics are determined by components external to the amplifier unit. The "differential" wording refers to the op amp input design where the output is determined by the difference of voltages between the two inputs. The ideal op amp offers infinite input impedance, zero output impedance, infinite gain, and a flat analog frequency response. **ANSWER D.**

G7B02 Which of the following is an advantage of using the binary system when processing digital signals?

A. Binary "ones" and "zeros" are easy to represent with an "on" or "off" state.

B. The binary number system is most accurate.

C. Binary numbers are more compatible with analog circuitry.

D. All of these choices are correct.

Now that high frequency transceivers employ numerous digital circuits, we need to review a little Boolean algebra. Oh, you missed that class?

A binary number 0 could result in your little handheld radio's LCD display showing a blank (OFF) segment. The binary number 1 may tell the LCD display to turn a segment ON. The reason your LCD handheld display draws such little current is that once each segment gets its command to be either OFF (0) or ON (1), it remains that way until it receives another binary command. Many digital logic circuits in your new HF transceiver provide command "paths" from one stage to another. Logic symbols give us a clue about what it takes to trigger the next ON or OFF sequence – and sometimes reversing them! Engineers write "Truth Tables" to program multiple inputs and a single output logic path. The most basic form of using the binary system for the processing of digital signals will lead to binary 0s triggering an OFF state and binary 1s triggering an ON state. *Just remember "binary," 1s and 0s, ONs or OFFs*. **ANSWER A.**

Elmer Point: *Here is a very important point for your understanding of digital electronics: To write and keep track of the many possible combinations of high or low signals in digital information, the combinations are treated as binary numbers. Binary numbers are also called base two numbers. An example is shown below.*

These days, most children learn about decimal numbers in elementary school. When written in ordinary decimal form, integers (whole numbers) are expressed as so many ones, so many tens, so many hundreds, and so forth. Each place in the number has a value ten times the place to the right. This requires using ten (for decimal) different symbols called decimal numerals or digits: 0, 1, 2, 3, 4, 5, 6, 7, 8, and 9. But in binary form, whole numbers are expressed as so many ones, so many twos, so many fours, and so forth. Each place in the number has twice the value of the next place to the right. Binary numbers use only two digits instead of ten: just 0 and 1.

So, in binary form, a number is written as a string of ones and zeroes. For instance, the integer one is written as 1. Two would be 10, which we read as "one-zero". It means a two and no one. Three would be 11 ("one-one", not eleven), meaning a two and a one. Four would be 100, called "one-zero-zero", meaning a four, no two, and no one. Five would be 101, meaning a four, no two, and a one. One hundred would be 1100100. That means a sixty-four, a thirty-two, no sixteen, no eight, a four, no two, and no one.

Each zero or one in a binary number is called a binary digit, or bit for short. Bit also means a little piece of information. In fact, a bit is the smallest possible piece of information in a digital system. It expresses a choice between only two possibilities. In a binary number, for instance, a particular bit can say either, "Yes, there is a 64 in this number," or, "No, there is not a 64 in this number."

ONE HUNDRED TWENTY-EIGHT
SIXTY-FOUR
THIRTY-TWO
SIXTEEN
EIGHT
FOUR
TWO
ONE

128
32
16
2

$$1\ 0\ 1\ 1\ 0\ 0\ 1\ 0 \quad = \quad \overline{178}$$

In digital systems, binary numbers are used to write and keep track of the many possible combinations of the two electrical states.

Source: *Basic Electronics* © 1994, Master Publishing, Inc., Niles, Illinois

Binary Number:

Each place has a value **twice as great** as the next place to the right. **Two numerals or digits** are used:

0 (Zero = "No")
1 (One = "Yes")

These digits are called bits.

Binary System

G7B03 Which of the following describes the function of a two input AND gate?

A. Output is high when either or both inputs are low.
B. Output is high only when both inputs are high.
C. Output is low when either or both inputs are high.
D. Output is low only when both inputs are high.

To tell you the truth, ham operators designing flip flop circuits will have their "truth table" for reference that allows them to view multiple flip flop configurations. If an input or output is "high," we represent it with the binary number ONE. If we look at a truth table, an *AND gate* will only give us a *high output when BOTH inputs are high*. **ANSWER B.**

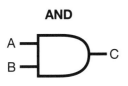

A	B	C
0	0	0
0	1	0
1	0	0
1	1	1

$C = A \bullet B$
$C = A B$

G7B04 Which of the following describes the function of a two input NOR gate?
A. Output is high when either or both inputs are low.
B. Output is high only when both inputs are high.
C. Output is low when either or both inputs are high.
D. Output is low only when both inputs are high.

Now they ask about the NOR gate. The two input *NOR gate* will give us a *LOW output when EITHER or BOTH inputs are high*. **ANSWER C.**

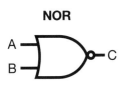

A	B	C
0	0	1
0	1	0
1	0	0
1	1	0

$C = \overline{A + B}$

G7B05 How many states does a 3-bit binary counter have?
A. 3.
B. 6.
C. 8.
D. 16.

Each flip-flop requires two input pulses to generate one output pulse. Each time you add another flip-flop in series, you multiple the number of inputs required to generate one output by a factor of two. $2 \times 2 \times 2 = 8$. **ANSWER C.**

G7B06 What is a shift register?
 A. A clocked array of circuits that passes data in steps along the array.
 B. An array of operational amplifiers used for tri state arithmetic operations.
 C. A digital mixer.
 D. An analog mixer.
A *shift register passes* flip-flop *data up or down* to each other, rather than just an output. Parallel data may be converted to serial data, or the reverse, through a shift register. **ANSWER A.**

G6C01 Which of the following is an analog integrated circuit?
 A. NAND Gate.   C. Frequency Counter.
 B. Microprocessor.   D. Linear voltage regulator.
Analog circuits offer a smooth ride down Grandpa's old house banister. Digital circuits are like bumping down individual steps. We find digital circuits in frequency counters, NAND gates, microprocessors, and any other digital circuit inside your new high frequency transceiver. Also inside this same set may be an *analog linear voltage regulator*, tied in with a reference diode and a pass transistor, regulating current to keep the desired voltage steady. No 0s and 1s here! Rather, a voltage error amplifier controls the current to the transistor providing increased or decreased conduction to maintain the desired voltage within the transceiver's specifications. **ANSWER D.**

G6C03 Which of the following is an advantage of CMOS integrated circuits compared to TTL integrated circuits?
 A. Low power consumption.
 B. High power handling capability.
 C. Better suited for RF amplification.
 D. Better suited for power supply regulation.
The big advantage of CMOS Logic over TTL is that *CMOS* offers *low power consumption*. **ANSWER A.**

G6C04 What is meant by the term ROM?
 A. Resistor Operated Memory.   C. Random Operational Memory.
 B. Read Only Memory.   D. Resistant to Overload Memory.
Your VHF/UHF handheld radio, and likely the modern high frequency radio you are ready to buy, contain *read only memory (ROM)*. It is permanent, and is programmed at the ham radio transceiver factory to set band limits, tuning ranges, and some other functions. This read only memory is permanent and does not need a battery backup. **ANSWER B.**

G6C05 What is meant when memory is characterized as "non-volatile"?
 A. It is resistant to radiation damage.
 B. It is resistant to high temperatures.
 C. The stored information is maintained even if power is removed.
 D. The stored information cannot be changed once written.
Another type of memory is called RAM – Random Access Memory. In some HF transceivers, a small button battery backs up all of those 100 channels of cool frequencies you have entered into random access memory. The internal battery needs to be changed every 5 years or so because the memory is volatile and will disappear if there is no voltage to keep it alive. *Non-volatile* RAM is what we see in newer transceivers, and *even though you pull the plug* and there is no internal

battery backup, information stored in this type of non-volatile *memory remains*.
ANSWER C.

G7B01 Complex digital circuitry can often be replaced by what type of integrated circuit?

A. Microcontroller.
B. Charge-coupled device.
C. Phase detector.
D. Window comparator.

You likely know that large scale integrated "chips" (LSIs) contain a microprocessor. There also are integrated circuits that will handle *complex digital circuitry*, called *microcontrollers*. **ANSWER A.**

An LSI mounted on a printed circuit board

Elmer Point: *Your Granddad ham likely knew that a chap named Georg Simon Ohm (1789-1854) experimented with electricity and discovered that the resistance (R) of a conductor depends on its length in feet, cross-sectional area in circular mils, and the resistivity, which is a parameter that depends on the molecular structure of the conductor and its temperature. Ohm's Law states:*

The current in an electrical circuit is directly proportional to the voltage and inversely proportional to the resistance.

The Ohm's Law and Power Circle shown here includes 12 equations that allow us to solve for voltage (E), current (I), resistance (R), and power (P), if we know the other values. There isn't much tough math in this section, but keep this page bookmarked to help you solve common electrical formulas.

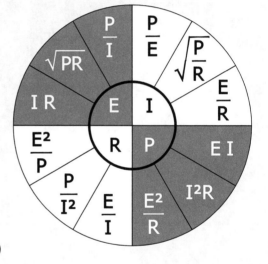

To solve for Voltage (E):

$E = I$ (current) $\times R$ (resistance)

$E = \sqrt{P \text{ (power)} \times R \text{ (resistance)}}$

$E = P$ (power) $\div I$ (current)

To solve for Current (I):

$I = P$ (power) $\div E$ (voltage)

$I = \sqrt{P \text{ (power)} \div R \text{ (resistance)}}$

$I = E$ (voltage) $\div R$ (resistance)

To solve for Resistance (R):

$R = E^2$ (voltage squared) $\div P$ (power)

$R = P$ (power) $\div I^2$ (current squared)

$R = E$ (voltage) $\div I$ (current)

To solve for Power (P):

$P = E$ (voltage) $\times I$ (current)

$P = I^2$ (current squared) $\times R$ (resistance)

$P = E^2$ (voltage squared) $\div R$ (resistance)

Source: The ARRL Handbook, 2207, © 2006, American Radio Relay League

G5B12 What would be the RMS voltage across a 50-ohm dummy load dissipating 1200 watts?

A. 173 volts.   C. 346 volts.
B. 245 volts.   D. 692 volts.

When you work with high-frequency mobile antennas or, in this question, a high-frequency dummy load, you may wish to calculate the voltage across the load to make sure things won't arc over, or to determine the maximum value of voltage that can be applied across the load without exceeding its power rating. Here is your formula:

$$E = \text{Square root of } P \times R = \sqrt{P \times R}$$
$$E = \text{Square root of } 50 \times 1200 = \sqrt{50 \times 1200}$$
$$E = \text{Square root of } 60{,}000 = \sqrt{60{,}000}$$
$$E = 244.948$$

This is an easy one to work out on a calculator – and yes, calculators are allowed in the exam room. Her are the keystrokes: Clear, Clear, 50 x 1200 = 60,000. Now simply tap the square root sign and, voila, the correct answer pops out at (approximately) 245 volts, which shows up as Answer B on your exam. Don't forget, examiners are allowed to scramble the A, B, C, D order, so simply look for *245 volts* as the correct answer. **ANSWER B.**

G5B03 How many watts of electrical power are used if 400 VDC is supplied to an 800-ohm load?

A. 0.5 watts.
B. 200 watts.
C. 400 watts.
D. 3200 watts.

The equation for this problem is $P = E^2 \div R$. 400^2 is 160,000. That divided by R (800) is 200, so the answer is *200 watts*. Calculator keystrokes are: Clear, 400 × 400 ÷ 800 = 200. **ANSWER B.**

G5B04 How many watts of electrical power are used by a 12-VDC light bulb that draws 0.2 amperes?

A. 2.4 watts.
B. 24 watts.
C. 6 watts.
D. 60 watts.

Easy equation – power is equal to voltage times current ($P = E \times I$). Multiply volts times amps, and you end up with *2.4 watts*. Using the magic circle for power, you see that power is equal to E × I. Calculator keystrokes are: Clear, 12 × 0.2 = 2.4 (in watts). **ANSWER A.**

G5B05 How many watts are dissipated when a current of 7.0 milliamperes flows through 1.25 kilohms?

A. Approximately 61 milliwatts.
B. Approximately 61 watts.
C. Approximately 11 milliwatts.
D. Approximately 11 watts.

Use the current and resistance version of the magic power circle above for the equation $P = I^2 \times R$. I = 0.007 amperes and R = 1250 ohms. Your answer will come out 0.061 watts. This is converted to *61 milliwatts* by moving the decimal point 3 places to the right. Calculator keystrokes are: Clear, 0.007 × 0.007 × 1250 = 0.06125 (in watts). **ANSWER A.**

G5B07 Which value of an AC signal results in the same power dissipation as a DC voltage of the same value?

A. The peak-to-peak value.
B. The peak value.
C. The RMS value.
D. The reciprocal of the RMS value.

Root Mean Square (RMS) measurement of an AC signal (also called the effective value of an AC voltage) is the same as a DC voltage of the same value. **ANSWER C.**

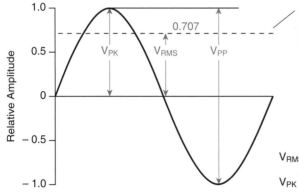

Root Mean Square Value
This value of AC voltage produces same heating in a resistor as a DC voltage of the same value.

$V_{RMS} = 0.707 \, V_{PK}$ $V_{PP} = 2 \times V_{PK}$

$V_{PK} = 1.414 \, V_{RMS}$ $V_{PK} = \dfrac{V_{PP}}{2}$

RMS (V_{RMS}), Peak, (V_{PK}), and Peak-to-Peak (V_{PP}) Voltage

G5B08 What is the peak-to-peak voltage of a sine wave that has an RMS voltage of 120 volts?

A. 84.8 volts.   C. 240.0 volts.
B. 169.7 volts.   D. 339.4 volts.

120 volts multiplied by 1.41 (average to peak) with the result multiplied by 2 (peak to peak) results in *339.4 volts*, which is what you would see on an oscilloscope sampling voltage out of your wall socket. You didn't realize your house wiring was that good, huh? **ANSWER D.**

G5B09 What is the RMS voltage of a sine wave with a value of 17 volts peak?

A. 8.5 volts.   C. 24 volts.
B. 12 volts.   D. 34 volts.

To go from 17 volts peak down to the RMS "average" voltage, multiply by 0.707. You end up with *12 volts* AC. **ANSWER B.**

G7A08 Which of the following is an advantage of a switch-mode power supply as compared to a linear power supply?

A. Faster switching time makes higher output voltage possible.
B. Fewer circuit components are required.
C. High frequency operation allows the use of smaller components.
D. All of these choices are correct.

The switching power supply is about the same price as the old, big, heavy transformer power supplies for 20 amps or 35 amps, 12-volt DC output. The switching power supply uses a relatively *high-frequency* oscillator at a frequency where *small, light weight, low-cost miniature transformers* create a relatively smooth DC power output. But one concern of the switching power supply is the proximity of it to your new high-frequency antenna system. Your antenna must be at least 10 feet or further away from the switching power supply so that it doesn't pick up broad-band noise from the switching power supply. **ANSWER C.**

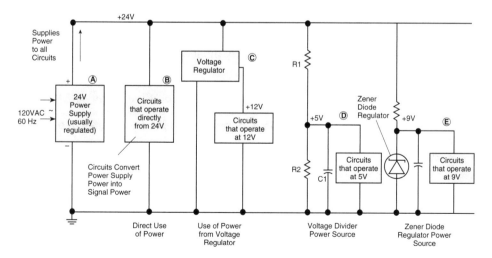

Regulated power supply provides power to the system in a variety of ways.

G6A01 Which of the following is an important characteristic for capacitors used to filter the DC output of a switching power supply?
A. Low equivalent series resistance.
B. High equivalent series resistance.
C. Low Temperature coefficient.
D. High Temperature coefficient.

Switching power supplies are those lightweight wonders that don't require a big, heavy transformer on the inside. Unlike the big transformer power supplies that use large electrolytic capacitors, *switching power supplies* employ *smaller capacitors* with *low equivalent series resistance* that make them more compatible with the frequency wave form switching techniques employed inside the switching supply. **ANSWER A.**

Modern switching power supplies incorporate "crowbar protection" to provide overvoltage protection

G6B04 When two or more diodes are connected in parallel to increase current handling capacity, what is the purpose of the resistor connected in series with each diode?
A. To ensure the thermal stability of the power supply.
B. To regulate the power supply output voltage.
C. To ensure that one diode doesn't carry most of the current.
D. To act as an inductor.

Adding a *resistor* in series with two or more diodes connected in parallel will help *balance*, equally, the *amount of current* that *each diode will pass*. **ANSWER C.**

G7A05 What portion of the AC cycle is converted to DC by a half-wave rectifier?
A. 90 degrees.   C. 270 degrees.
B. 180 degrees.   D. 360 degrees.

The half-wave rectifier uses only half of the cycle, which is *180 degrees*. **ANSWER B.**

G7A06 What portion of the AC cycle is converted to DC by a full-wave rectifier?
A. 90 degrees.   C. 270 degrees.
B. 180 degrees.   D. 360 degrees.

A full-wave rectifier is much more efficient because it uses all *360 degrees* of the AC cycle. A full-wave rectifier output also is much easier to filter to provide pure DC voltage. **ANSWER D.**

G7A07 What is the output waveform of an unfiltered full-wave rectifier connected to a resistive load?
    A. A series of DC pulses at twice the frequency of the AC input.
    B. A series of DC pulses at the same frequency as the AC input.
    C. A sine wave at half the frequency of the AC input.
    D. A steady DC voltage.
A full-wave rectifier gives a much smoother *pulsating DC* to filter than a half-wave rectifier because the full-wave rectified half sine waves are *double the line frequency*. **ANSWER A.**

G6B01 What is the peak-inverse-voltage rating of a rectifier?
    A. The maximum voltage the rectifier will handle in the conducting direction.
    B. 1.4 times the AC frequency.
    C. The maximum voltage the rectifier will handle in the non-conducting direction.
    D. 2.8 times the AC frequency.
The peak-inverse-voltage (PIV) rating of a power supply rectifier is the *maximum voltage it will handle in the non-conducting direction*. **ANSWER C.**

G7A03 What is the peak-inverse-voltage across the rectifiers in a full-wave bridge power supply?
    A. One-quarter the normal output voltage of the power supply.
    B. Half the normal output voltage of the power supply.
    C. Double the normal peak output voltage of the power supply.
    D. Equal to the normal peak output voltage of the power supply.
The full-wave bridge rectifier circuit offers an output of pulsating DC that is far easier to smooth out than either a single-diode half-wave circuit or a two-diode full-wave center tap rectifier circuit. This full wave bridge develops a Peak Inverse Voltage across the 4 rectifier diodes that is nearly *equal to the normal peak output of the power supply*. **ANSWER D.**

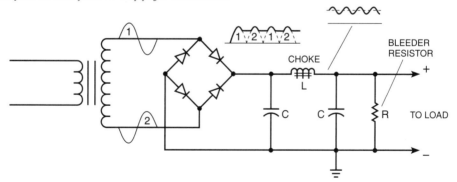

Full-Wave Power Supply with Bleeder Resistor

G7A04 What is the peak-inverse-voltage across the rectifier in a half-wave power supply?
    A. One-half the normal peak output voltage of the power supply.
    B. One-half the normal output voltage of the power supply.
    C. Equal to the normal output voltage of the power supply.
    D. Two times the normal peak output voltage of the power supply.

In the half-wave rectifier, the voltage across that single hard-working diode is two times the normal peak output voltage, because the smoothing output capacitor will hold the peak voltage during the negative half of the cycle, while the transformer will be at the negative peak, letting this single diode see *twice the normal peak output voltage* across it! **ANSWER D.**

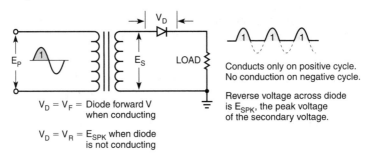

$V_D = V_F =$ Diode forward V when conducting

$V_D = V_R = E_{SPK}$ when diode is not conducting

Conducts only on positive cycle. No conduction on negative cycle.

Reverse voltage across diode is E_{SPK}, the peak voltage of the secondary voltage.

Half-Wave Rectifier

G6B02 What are two major ratings that must not be exceeded for silicon diode rectifiers?
 A. Peak inverse voltage; average forward current.
 B. Average power; average voltage.
 C. Capacitive reactance; avalanche voltage.
 D. Peak load impedance; peak voltage.
Be sure to never exceed the *peak-inverse-voltage* rating and the average *forward current* rating of silicon-diode rectifiers used in power supplies. **ANSWER A.**

G7A02 Which of the following components are used in a power-supply filter network?
 A. Diodes.   C. Quartz crystals.
 B. Transformers and transducers.   D. Capacitors and inductors.
In power supplies, transformers supply the voltage and current, diodes rectify, and *capacitors and inductors filter*. A bleeder resistor protects. **ANSWER D.**

G6A04 Which of the following is an advantage of an electrolytic capacitor?
 A. Tight tolerance.   C. High capacitance for given volume.
 B. Non-polarized.   D. Inexpensive RF capacitor.
We usually find the electrolytic capacitor in the power supply section of our new rig. The electrolytic capacitor is polarized with a positive and negative connection point. The electrolytic offers *high capacitance for given volume* (volume = size). **ANSWER C.**

G6A02 Which of the following types of capacitors are often used in power supply circuits to filter the rectified AC?
 A. Disc ceramic.   C. Mica.
 B. Vacuum variable.   D. Electrolytic.
Rectified AC is a form of pulsating DC. Electrolytic capacitors usually are used as filters to smooth pulsating DC because they offer large amounts of capacity in a small size. The big problem with *electrolytic capacitors* – especially old ones – is that they dry out and lose their capacitance. **ANSWER D.**

G6A12 What is a common name for an inductor used to help smooth the DC output from the rectifier in a conventional power supply?

A. Back EMF choke.   C. Charging inductor.
B. Repulsion coil.   D. Filter choke.

The half wave rectifier will produce a small amount of ripple identical to the line input frequency, normally 60 Hz. Full wave rectifiers will produce twice the 60 Hz ripple, or 120 Hz. The *filter choke inductor* is much more efficient at this double AC frequency to smooth out the AC ripple imposed on the DC line. **ANSWER D.**

G7A01 What safety feature does a power-supply bleeder resistor provide?

A. It acts as a fuse for excess voltage.
B. It discharges the filter capacitors.
C. It removes shock hazards from the induction coils.
D. It eliminates ground-loop current.

When you turn off your big rig, it dims down and then cycles off completely. This slow decay of voltage is from the *filter capacitors* that are slowly *being discharged* by the bleeder resistors. This is a safety feature, and also helps provide voltage regulation. See the illustration at G7A03, page 125. **ANSWER B.**

G4B06 What is an advantage of a digital voltmeter as compared to an analog voltmeter?

A. Better for measuring computer circuits.
B. Better for RF measurements.
C. Better precision for most uses.
D. Faster response.

Your buddy discovers you are a brand new General and gives you several gel-cell 12 volt batteries. To see which one has the absolute best charge, use a digital volt meter to get a *more precise digital readout* of the battery voltage. A neat little analog volt meter might not show you subtle changes in battery terminal voltage. **ANSWER C.**

G4B14 What is an instance in which the use of an instrument with analog readout may be preferred over an instrument with a numerical digital readout?

A. When testing logic circuits.
B. When high precision is desired.
C. When measuring the frequency of an oscillator.
D. When adjusting tuned circuits.

Don't throw away that old Simpson model 260 needle meter. The analog needle movement can give you the "feel," visually, of *tuned circuits*, as long as that particular tuned circuit does not call for specialized, non-loading measurements. That old needle readout may also be less affected by strong nearby signals up at the repeater site, too! **ANSWER D.**

Analog multimeter.

G4B05 Why is high input impedance desirable for a voltmeter?
A. It improves the frequency response.
B. It decreases battery consumption in the meter.
C. It improves the resolution of the readings.
D. It decreases the loading on circuits being measured.
New digital volt meters have a relatively *high input impedance to avoid loading* down *the circuit being measured*. If you are measuring a fraction of a volt, you don't want to use an older volt meter that might significantly add load to the circuit and result in a bogus reading. **ANSWER D.**

G0B16 When might a lead acid storage battery give off explosive hydrogen gas?
A. When stored for long periods of time.
B. When being discharged.
C. When being charged.
D. When not placed on a level surface.
I run my home station off of a lead acid storage battery to keep me on the air in case of a blackout. The battery is outside because it gives off *explosive hydrogen gas when it is being charged* by solar panels, and it works very well. Never use a storage battery inside your ham shack because of the danger of explosive hydrogen gas. **ANSWER C.**

Large storage battery.

G6B14 What is the minimum allowable discharge voltage for maximum life of a standard 12 volt lead acid battery?
A. 6 volts.   C. 10.5 volts.
B. 8.5 volts.   D. 12 volts.
When operating your equipment with a 12 volt lead acid automobile battery during Field Day, make sure you never pull your battery below *10.5 volts*. If you do, you will shorten the maximum life of that battery, and any transmitter running at 10.5 volts (instead of 12 volts) will usually sound distorted on high frequency. **ANSWER C.**

G6B15 When is it acceptable to recharge a carbon-zinc primary cell?
A. As long as the voltage has not been allowed to drop below 1.0 volt.
B. When the cell is kept warm during the recharging period.
C. When a constant current charger is used.
D. Never.
Disposable flashlight "D" cells are *never* intended for *recharge*. Attempts to recharge carbon zinc primary cells or newer alkaline batteries may lead to the battery venting dangerous gas. Unless the battery specifically states "rechargeable," do NOT try to recharge NON-rechargeable cells. **ANSWER D.**

G6B13 What is an advantage of the low internal resistance of nickel-cadmium batteries?
A. Long life.   C. High voltage.
B. High discharge current.   D. Rapid recharge.

The Nickel Cadmium battery is a low-cost, rechargeable voltage source for handheld radios as well as for some QRP (low power output) high frequency transceivers. The low internal resistance of the nickel cadmium battery allows for *high discharge current* when transmitting. The disadvantage of low internal resistance is the slight self-discharge between uses of your radio equipment. So, before going out on Field Day, be sure to cycle your nickel cadmium battery pack several times, and end the cycle with a good charge. **ANSWER B.**

G4E08 What is the name of the process by which sunlight is changed directly into electricity?

A. Photovoltaic conversion.

B. Photon emission.

C. Photosynthesis.

D. Photon decomposition.

You can change sunlight into voltage, and notice the word "volt" in the term *photovoltaic conversion*. **ANSWER A.**

Solar Panel array for
Charging Storage Batteries

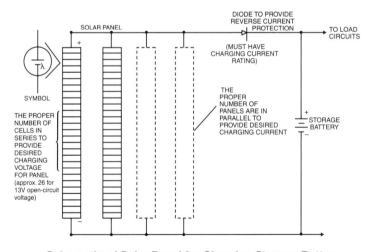

Schematic of Solar Panel for Charging Storage Battery

G4E09 What is the approximate open-circuit voltage from a modern, well-illuminated photovoltaic cell?

A. 0.02 VDC.

B. 0.5 VDC.

C. 0.2 VDC.

D. 1.38 VDC.

A complete photovoltaic cell will yield *0.5 volts* direct current. A solar panel is made up of a series-parallel connection of these cells in order to charge your storage battery system. **ANSWER B.**

G4E10 What is the reason a series diode is connected between a solar panel and a storage battery that is being charged by the panel?
    A. The diode serves to regulate the charging voltage to prevent overcharge.
    B. The diode prevents self discharge of the battery though the panel during times of low or no illumination.
    C. The diode limits the current flowing from the panel to a safe value.
    D. The diode greatly increases the efficiency during times of high illumination.
I always like to monitor solar power charging with a small amp meter in series with the red lead. You can really see how a single shadow will decrease output. Then one night you check the panel and its actually showing a discharge on your battery. There probably was no series diode to prevent the panel from slightly discharging your battery. Newer panels usually have the diodes in place. But some older panels, or scavenged panels without a controller, could really use that *reverse current diode to prevent battery discharge*. **ANSWER B.**

G4E11 Which of the following is a disadvantage of using wind as the primary source of power for an emergency station?
    A. The conversion efficiency from mechanical energy to electrical energy is less than 2 percent.
    B. The voltage and current ratings of such systems are not compatible with amateur equipment.
    C. A large energy storage system is needed to supply power when the wind is not blowing.
    D. All of these choices are correct.
The wind doesn't blow all the time, so wind power is not a good primary source for your emergency communications station. When the wind isn't blowing, you need a *huge bank of batteries* to keep your station on the air. **ANSWER C.**

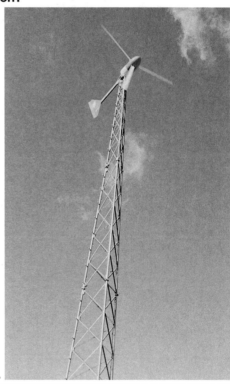

No wind – no juice.

G4E03 Which of the following direct, fused power connections would be the best for a 100-watt HF mobile installation?
    A. To the battery using heavy gauge wire.
    B. To the alternator or generator using heavy gauge wire.
    C. To the battery using resistor wire.
    D. To the alternator or generator using resistor wire.

As a General Class operator, you will probably be running a 100-watt, HF mobile transceiver in your car. You cannot rely on the car's 12-volt accessory or "power port" socket wiring to support the necessary 20-amp (minimum) power demands from your new radio. Wire your red and black power leads *directly to the battery using heavy-gauge wire*. Fuse both the red and the black power leads close to the battery connection point. **ANSWER A.**

Elmer Point: *To go mobile with your 100-watt HF ham transceiver, you'll need to run the red and black power wires directly to the positive and negative terminals on your car or truck battery. You should have separate fuses right next to the battery terminal connections for safety. While automotive sound systems sometimes use the vehicle chassis as the negative black wire return, commercial two-way radio installers always say it's best to run directly to the positive and negative terminals on the battery. The chassis of your HF radio also should be well grounded to your vehicle frame.*

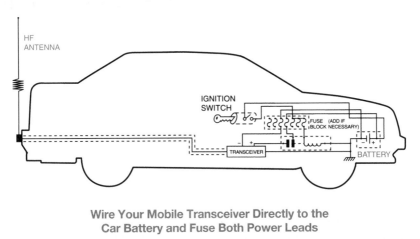

Wire Your Mobile Transceiver Directly to the Car Battery and Fuse Both Power Leads

G4E04 Why is it best NOT to draw the DC power for a 100-watt HF transceiver from an automobile's auxiliary power socket?

A. The socket is not wired with an RF-shielded power cable.
B. The socket's wiring may be inadequate for the current being drawn by the transceiver.
C. The DC polarity of the socket is reversed from the polarity of modern HF transceivers.
D. Drawing more than 50 watts from this socket could cause the engine to overheat.

While you might be tempted to grab 12 volts from an automobile auxiliary power socket, DON'T. Sure, the radio will work for a few minutes, but after a little bit of transmitting the auxiliary power receptacle wiring will get red hot from *over-drawing current*, and quite possibly cause a fire in your dashboard! **ANSWER B.**

G6C15 What is the main reason to use keyed connectors instead of non-keyed types?

A. Prevention of use by unauthorized persons.

B. Reduced chance of incorrect mating.

C. Higher current carrying capacity.

D. All of these choices are correct.

Metal microphone plugs and jacks all have a protruding ridge or a channel so that the metal microphone plug fits properly into the receptacle. They call this a "keyed" connector and it *reduces the chance* of accidentally *bending fragile pins* when you go to plug in your microphone connector. **ANSWER B.**

Keyed Connector.

Elmer Point: *Want to learn more about electricity and how electronics work? Here are two books I recommend highly for your self-education!*

Getting Started in Electronics *by Forrest M. Mims III is a true classic. It is used by a wide range of people – from junior-high teachers to the U.S. Army to teach the fundamentals of electricity and electronics.*

Basic Electronics *by Alvis J. Evans and Gene McWhorter goes a little deeper into the topic, and includes end of chapter quizzes and worked-out problems to teach you in detail the various aspects of electronics.*

Either book – or both – will give you a solid grounding in the theory, science, and practical applications of electronics. You can get your copies of the books at your local ham radio store, on line at www.w5yi.org, or by calling The W5YI Group at 800-669-9594.

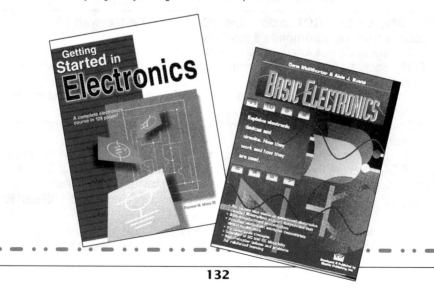

Website Resources

▼ IF YOU'RE LOOKING FOR	▼ THEN VISIT
equipment reviews	www.hamoperator.com
Call sign lookups and more	www.qrz.com
ham radio gear, call sign look-ups	www.hamcall.net
QSO's and more	www.hamgallery.com
ham radio resources	www.dxzone.com
articles, reviews, etc.	www.eham.net
used equipment ads and more	www.qth.com
operating tips from Calgary hams	www.cara.ampr.org
directional antennas	www.arrowantennas.com
direction finding tips and more	www.homingin.com
free e-mail service for hams	www.qsl.net
ham radio for people with disabilities	www.handiham.org
Canada's amateur radio society	www.rac.ca
operating tips, ham news	www.hamquick.com
FCC amateur radio enforcement log	www.rainreport.com
California club with good tech articles	www.cvarc.org/faq.htm
antennas and related gear	www.natcommgroup.com
every ham accessory known to man	www.mfjenterprises.com
CQ Magazine's website	www.cq-amateur-radio.com
radio propagation reports	www.dxworld.com/50prop.html
news & science reports about space	www.spacetoday.org
mobile antennas and accessories	www.hiqantennas.com
educational resources and links	www.ecjones.org
antennas, connectors, and more	www.cq73.com
linking ham radio via the internet	www.winlink.org
inductive components for RFI	www.amidoncorp.com
HF amplifiers, antennas, and more	www.ameritron.com
feedlines, connectors, & wire galore	www.cablexperts.com
copper ground strap, antenna masts	www.metal-cable.com
all kinds of antennas & accessories	www.antennaworld.com
antenna site with good technical data	www.cushcraft.com
excellent RF safety calculator	http://n5xu.ae.utexas.edu/rfsafety
technical resources for hams	www.csvhfs.org
old instruction manuals	http://bama.sbc.edu
RF safety calculator	http://hintlink.com/power_density.htm

Circuits

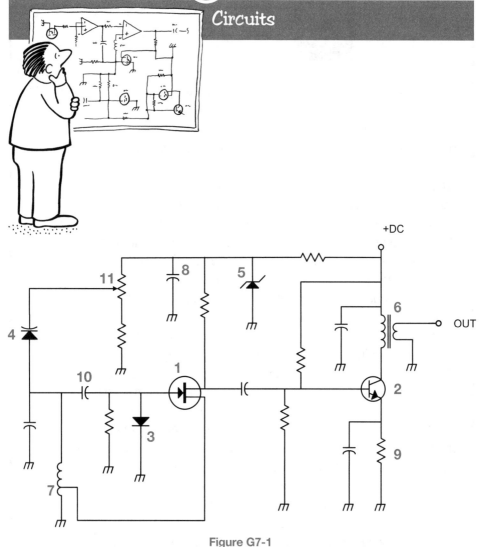

Figure G7-1

Look at Figure G7-1. On your exam, it may appear on the last page of the examination sheets, or may appear for a single question about the components. We can identify this circuit as a Hartley oscillator, with symbol 7 as the telltale tapped coil. This is a voltage-controlled Hartley oscillator, seen by the varactor, symbol 4. Symbol 11 is a variable potentiometer, and symbol 8 is a bypass capacitor. Symbol 10 is a blocking capacitor, isolating out DC. Symbol 1 is a field effect transistor (JFET) with symbol 5 being a Zener diode for voltage regulation. Symbol 6 is the oscillator's companion amplifier output transformer, and symbol 2 is an NPN transistor. Symbol 9 is a fixed resistor, and symbol 3 is a diode. Now, let's see what they're going to ask on the exam.

G7A09 Which symbol in figure G7-1 represents a field effect transistor?
    A. Symbol 2.   C. Symbol 1.
    B. Symbol 5.   D. Symbol 4.
Symbol 1 is the Field Effect Transistor. Notice the arrow is NOT
POINTING IN, so this is an N Channel FET. **ANSWER C.**

Symbol 1

G7A10 Which symbol in figure G7-1 represents a Zener diode?
    A. Symbol 4.   C. Symbol 11.
    B. Symbol 1.   D. Symbol 5.
Symbol 5 is the Zener diode, used for voltage regulation. **ANSWER D.**

Symbol 5

G7A11 Which symbol in figure G7-1 represents an NPN junction transistor?
    A. Symbol 1.   C. Symbol 7.
    B. Symbol 2.   D. Symbol 11.
See the transistor all the way to the right, *symbol 2*? Again, the arrow is NOT
POINTING IN so it is an NPN junction transistor. **ANSWER B.**

Symbol 2

G7A12 Which symbol in Figure G7-1 represents a multiple-winding transformer?
    A. Symbol 4.   C. Symbol 6.
    B. Symbol 7.   D. Symbol 1.
It's easy to spot the transformer – we find it at *symbol 6*, and the 2
vertical lines may indicate an iron core. **ANSWER C.**

Symbol 6

G7A13 Which symbol in Figure G7-1 represents a tapped inductor?
    A. Symbol 7.   C. Symbol 6.
    B. Symbol 11.   D. Symbol 1.
This is what gives away the Hartley oscillator; *symbol 7*, the tapped
inductor. This provides us with the feedback necessary to keep the
oscillator oscillating!
ANSWER A.

Symbol 7

Elmer Point: *Electrical circuits are represented graphically in schematic diagrams using symbols to indicate the various components. Here's a table of schematic symbols that will help you read the diagram used for questions G7A09 through G7A13. If you really get into experimental electronics or kit building, you might want to buy a copy of* Forrest Mims's **Engineer's Mini Notebook #4, Electronic Formulas, Symbols, & Circuits** *available at your local ham radio store, on-line at www.w5yi.org or by calling the W5YI Group at 800-669-9594.*

Volume IV

Electronic Formulas, Symbols & Circuits

ENGINEER'S Mini Notebook

by Forrest M. Mims III
www.forrestmims.com

Forrest M. Mims III

Hundreds of Electronics References:
- Formulas
- Tables
- Circuit Symbols
- Device Packages
- Design Tips
- Testing Tips

Includes Many Circuits Using:
- Diodes
- Transistors
- Power FETs
- Piezoelectric Buzzers

Basic Logic & Digital Circuits:
- TTL
- CMOS
- Interfacing

Master Publishing, Inc.

SCHEMATIC SYMBOLS

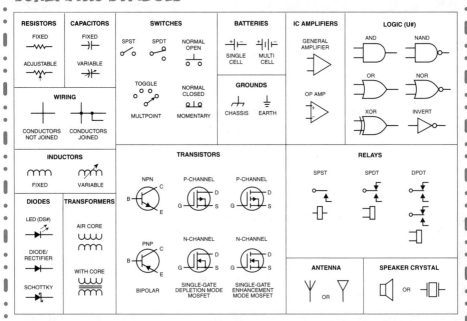

G5C04 What is the total resistance of three 100-ohm resistors in parallel?

A. 0.30 ohms.   C. 33.3 ohms.
B. 0.33 ohms.   D. 300 ohms.

If we have three equal value resistors in parallel, we will have three individual paths for current to flow, decreasing each like resistor's ohmic resistance by 1/3. You can do this one in your head: 1/3 the resistance of each 100 ohm resistor is *33.3 ohms*. If you want to do it the long way, here is the formula:

$R_T = 1 \div (1/R_1 + 1/R_2 + 1/R_3)$ **ANSWER C.**

Elmer Point: *Here is a set of questions that asks about the total value of resistors, capacitors, and inductors connected in SERIES and in PARALLEL. Resistors (R) and Inductors (L) act the same way. Capacitors (C) act in the opposite way. Here are the formulas to calculate the total value of these components:*

$$\text{Resistors in SERIES simply add up:} \quad R_{total} = R_1 + R_2 + R_3$$
$$\text{Inductors in SERIES simply add up:} \quad L_{total} = L_1 + L_2 + L_3$$

Resistors and Inductors in PARALLEL combine with a resulting total value that is always LESS than the value of the lowest value resistor in parallel, for easy problem solving! Here are the formulas:

$$\text{when there are 2 resistors or inductors:} \quad R_{total} = \frac{R_1 \times R_2}{R_1 + R_2} \quad or \quad L_{total} = \frac{L_1 \times L_2}{L_1 + L_2}$$

$$\text{when there are 3 or more resistors:} \quad R_{total} = \frac{1}{\dfrac{1}{R_1} + \dfrac{1}{R_2} + \dfrac{1}{R_3}} \quad or \quad L_{total} = \frac{1}{\dfrac{1}{L_1} + \dfrac{1}{L_2} + \dfrac{1}{L_3}}$$

Capacitors in PARALLEL simply add up: $C_{total} = C_1 + C_2 + C_3$

Capacitors in SERIES combine with a resulting total value that is always LESS than the value of the lowest value capacitor in series, for easy problem solving! Here is the formula:

$$\text{when there are 2 capacitors:} \quad C_{total} = \frac{C_1 \times C_2}{C_1 + C_2}$$

$$\text{when there are 3 or more capacitors:} \quad C_{total} = \frac{1}{\dfrac{1}{C_1} + \dfrac{1}{C_2} + \dfrac{1}{C_3}}$$

G6A06 What will happen to the resistance if the temperature of a resistor is increased?

A. It will change depending on the resistor's reactance coefficient.
B. It will stay the same.
C. It will change depending on the resistor's temperature coefficient.
D. It will become time dependent.

Heating a resistor always decreases its resistance. The amount of change for any particular temperature change depends on the *resistor's temperature coefficient*, which depends on the materials used in the resistor's construction. **ANSWER C.**

G6A08 Which of the following describes a thermistor?
A. A resistor that is resistant to changes in value with temperature variations.
B. A device having a specific change in resistance with temperature variations.
C. A special type of transistor for use at very cold temperatures.
D. A capacitor that changes value with temperature.
When you earn your new General license, you may hook up with another station thousands of miles away who will ask you what the weather is. Likely you have one of those new wireless weather stations, and the temperature sensor uses a component called a thermistor. The *thermistor* is a *resistor* that is designed to maximize *changes in value with temperature variations*. This makes it a great reference for slight changes in temperature. **ANSWER B.**

G5C10 What is the inductance of three 10 millihenry inductors connected in parallel?
A. 0.30 Henrys.   C. 3.3 millihenrys.
B. 3.3 Henrys.   D. 30 millihenrys.
Treat inductors like resistors when working an either series or parallel problem. This one you can do in your head. Three 10 millihenry inductors in parallel, total inductance will be 1/3 or *3.3 millihenrys* when connected in parallel. **ANSWER C.**

G5B02 How does the total current relate to the individual currents in each branch of a parallel circuit?
A. It equals the average of each branch current.
B. It decreases as more parallel branches are added to the circuit.
C. It equals the sum of the currents through each branch.
D. It is the sum of the reciprocal of each individual voltage drop.
If you *add up the current* in each branch of a parallel circuit, you will come up with the total current in the circuit. **ANSWER C.**

G5C05 If three equal value resistors in parallel produce 50 ohms of resistance, and the same three resistors in series produce 450 ohms, what is the value of each resistor?
A. 1500 ohms.   C. 150 ohms.
B. 90 ohms.   D. 175 ohms.
If we have three like resistors in parallel that together produce 50 ohms of resistance, each individual resistor would have a value of 150 ohms (50+50+50). Now check your answer – if we have three *150 ohm* resistors in series, the total resistance adds up to 450 ohms. **ANSWER C.**

G5C15 What is the total resistance of a 10 ohm, a 20 ohm, and a 50 ohm resistor in parallel?
A. 5.9 ohms.   C. 10000 ohms.
B. 0.17 ohms.   D. 80 ohms.
While you can solve this problem with a big formula, always remember that with unlike resistors in parallel, just like unlike capacitors in series, the resulting answer will always be less than the smallest value component. *5.9 ohms* is a logical answer to solve for when you work the big long formula all the way out. **ANSWER A.**
$$R_T = 1 \div (1/R_1 + 1/R_2 + 1/R_3)$$

G5C03 Which of the following components should be added to an existing resistor to increase the resistance?
 A. A resistor in parallel.   C. A capacitor in series.
 B. A resistor in series.   D. A capacitor in parallel.
Another easy one here – to add more resistance to a circuit, we *add resistor(s) in series*. **ANSWER B.**

G5C08 What is the equivalent capacitance of two 5000 picofarad capacitors and one 750 picofarad capacitor connected in parallel?
 A. 576.9 picofarads.   C. 3583 picofarads.
 B. 1733 picofarads.   D. 10750 picofarads.
Calculating total capacitance in parallel is easy – it is the sum of each individual capacitor. 5000 + 5000 (remember they say two) + 750 = *10750 picofarads*.
Capacitors in parallel simply add up. The formula is: $C_T = C_1 + C_2 + C_3$.
ANSWER D.

G5C09 What is the capacitance of three 100 microfarad capacitors connected in series?
 A. 0.30 microfarads.   C. 33.3 microfarads.
 B. 0.33 microfarads.   D. 300 microfarads.
We have three like capacitors in series – since they are in series, like resistors in parallel, total capacitance will be 1/3 of each 100 microfarad cap. *33.3 microfarads* is an answer you can do in your head! The formula is:
$C_T = 1 \div (1/C_1 + 1/C_2 + 1/C_3.)$ **ANSWER C.**

G5C11 What is the inductance of a 20 millihenry inductor in series with a 50 millihenry inductor?
 A. 0.07 millihenrys.   C. 70 millihenrys.
 B. 14.3 millihenrys.   D. 1000 millihenrys.
Easy one. Just like resistors add up in series, so do inductors. 20 + 50 = *70 millihenrys*. **ANSWER C.**

G5C12 What is the capacitance of a 20 microfarad capacitor in series with a 50 microfarad capacitor?
 A. 0.07 microfarads.   C. 70 microfarads.
 B. 14.3 microfarads.   D. 1000 microfarads.
Calculating for total capacitance of unlike capacitors in series is much like the formula for calculating unlike resistors in parallel: $(C_1 \times C_2) \div (C_1 + C_2)$. In this equation, the total capacitance will always be less than the smaller capacitor, so *14.3 microfarads* can be confirmed as the correct answer. **ANSWER B.**

G5C13 Which of the following components should be added to a capacitor to increase the capacitance?
 A. An inductor in series.   C. A capacitor in parallel.
 B. A resistor in series.   D. A capacitor in series.
If we need to add some additional capacitance to a circuit which already has a fixed capacitor, we would add a second *capacitor in PARALLEL*. **ANSWER C.**

G5C14 **Which of the following components should be added to an inductor to increase the inductance?**

A. A capacitor in series.   C. An inductor in parallel.
B. A resistor in parallel.   D. An inductor in series.

If we need to increase the inductance (L) of a circuit, we would simply add another *inductor in SERIES.* **ANSWER D.**

G5A02 **What is reactance?**

A. Opposition to the flow of direct current caused by resistance.
B. Opposition to the flow of alternating current caused by capacitance or inductance.
C. A property of ideal resistors in AC circuits.
D. A large spark produced at switch contacts when an inductor is de-energized.

Inductive *reactance is the opposition to AC* caused by inductors. Capacitive reactance is the opposition to AC caused by capacitors. Both reactances vary with frequency. When there is an inductor and a capacitor in the same circuit, there is a special frequency, called the resonant frequency, where the inductive reactance equals the capacitive reactance. **ANSWER B.**

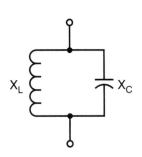

$$X_L = 2\pi f L$$

$$X_C = \frac{1}{2\pi f C}$$

The resonant frequency of a circuit is:

$$f_r = \frac{1}{2\pi \sqrt{LC}}$$

The resonant frequency is the frequency where $X_L = X_C$.

$$\therefore 2\pi f L = \frac{1}{2\pi f C}$$

$$f^2 = \frac{1}{(2\pi L)(2\pi C)}$$

$$f^2 = \frac{1}{(2\pi)^2 LC}$$

$$\therefore f_r = \frac{1}{2\pi \sqrt{LC}}$$

Resonant Frequency

G5A03 **Which of the following causes opposition to the flow of alternating current in an inductor?**

A. Conductance.   C. Admittance.
B. Reluctance.   D. Reactance.

Think of an inductor as a coil of wire. Its opposition to AC is called inductive *reactance*, identified as X_L. $X_L = 2\pi f L$ where f is the frequency in hertz and L is the inductance in henries. X_L increases as frequency increases. **ANSWER D.**

G5A09 **What unit is used to measure reactance?**

A. Farad.   C. Ampere.
B. Ohm.   D. Siemens.

The *ohm* is the unit of measurement for reactance as well as resistance. **ANSWER B.**

G5A04 Which of the following causes opposition to the flow of alternating current in a capacitor?

A. Conductance.
B. Reluctance.
C. Reactance.
D. Admittance.

A capacitor has plates separated by an insulating dielectric. Its opposition to AC is called capacitive *reactance*, identified as X_c. $X_c = 1 \div 2\pi fC$ where f is the frequency in hertz and C is the capacitance in farads. X_c decreases as frequency increases. **ANSWER C.**

G5A06 How does a capacitor react to AC?

A. As the frequency of the applied AC increases, the reactance decreases.
B. As the frequency of the applied AC increases, the reactance increases.
C. As the amplitude of the applied AC increases, the reactance increases.
D. As the amplitude of the applied AC increases, the reactance decreases.

Capacitors offer reactance to AC inversely proportional to the frequency. Capacitors have high reactance at low frequencies and *low reactance at high frequencies*. Remember, as f increases, X_c decreases ($X_c = 1 \div 2\pi fC$). **ANSWER A.**

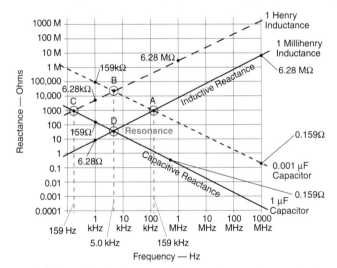

Variation of inductance and capacitive reactance with frequency (illustration not to exact log-log scale).
Source: *Basic Communications Electronics,* © 1999 Master Publishing, Inc., Niles, IL

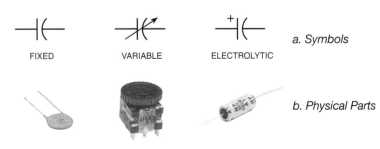

FIXED   VARIABLE   ELECTROLYTIC   *a. Symbols*

b. Physical Parts

Capacitors

G5A05 How does an inductor react to AC?
   A. As the frequency of the applied AC increases, the reactance decreases.
   B. As the amplitude of the applied AC increases, the reactance increases.
   C. As the amplitude of the applied AC increases, the reactance decreases.
   D. As the frequency of the applied AC increases, the reactance increases.

Inductors (coils) are effective in reducing alternator whine in high-frequency mobile installations. The *higher* the alternator whine *frequency*, the *greater* the *reactance* from the inductor. Remember, as f increases, X_L increases ($X_L = 2\pi fL$).
ANSWER D.

G5A01 What is impedance?
   A. The electric charge stored by a capacitor.
   B. The inverse of resistance.
   C. The opposition to the flow of current in an AC circuit.
   D. The force of repulsion between two similar electric fields.

The term *impedance* means the *opposition to the flow of alternating current in a circuit*. Impedance to AC can be made up of resistance only, reactance only, or both resistance and reactance. You can create impedance to AC by winding a wire around a pencil to create a coil. This handy "choke" might minimize the alternator whine that may come in on your new worldwide mobile high-frequency station temporarily mounted in your vehicle. **ANSWER C.**

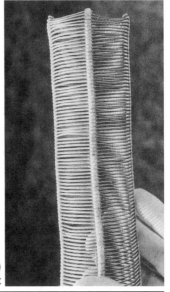

Coils create impedance (opposition)
to the flow of AC in a circuit

G5A10 What unit is used to measure impedance?
   A. Volt.   C. Ampere.
   B. Ohm.   D. Watt.

The *ohm* is also used for measuring impedance. Thus, the ohm may mean impedance, reactance, or resistance. **ANSWER B.**

G5A11 Which of the following describes one method of impedance matching between two AC circuits?
   A. Insert an LC network between the two circuits.
   B. Reduce the power output of the first circuit.
   C. Increase the power output of the first circuit.
   D. Insert a circulator between the two circuits.

Two AC circuits might be impedance matched by *using coils and capacitors (LC) between the two circuits*. **ANSWER A.**

G5A12 What is one reason to use an impedance matching transformer?
A. To minimize transmitter power output.
B. To maximize the transfer of power.
C. To reduce power supply ripple.
D. To minimize radiation resistance.

When impedances are matched, we will have the greatest amount of power transfer. An *impedance matching transformer* allows us to precisely match radio stages for the *maximum transfer of power*. **ANSWER B.**

G5A08 Why is impedance matching important?
A. So the source can deliver maximum power to the load.
B. So the load will draw minimum power from the source.
C. To ensure that there is less resistance than reactance in the circuit.
D. To ensure that the resistance and reactance in the circuit are equal.

When internal *source and load impedances* are *matched, maximum power* will be delivered to the load. Most new ham HF radios will automatically reduce power output when there is an impedance mismatch. **ANSWER A.**

G5A13 Which of the following devices can be used for impedance matching at radio frequencies?
A. A transformer.
B. A Pi-network.
C. A length of transmission line.
D. All of these choices are correct.

There are plenty of ways we can match impedances at radio frequencies. Up at the antenna, we sometimes will use fractional wavelength impedance-matching transmission lines. In a radio RF output stage, we might use an impedance-matching Pi-network. And within the radio, small transformers will allow us to impedance match. *All of these* are great ways for providing the maximum transfer of radio frequency energy. **ANSWER D.**

G5A07 What happens when the impedance of an electrical load is equal to the internal impedance of the power source?
A. The source delivers minimum power to the load.
B. The electrical load is shorted.
C. No current can flow through the circuit.
D. The source can deliver maximum power to the load.

Always make sure your new General Class worldwide antenna systems have an impedance around 50 ohms for *maximum power transfer*. Some General Class ham transceivers have built-in, automatic impedance-matching antenna tuner networks. **ANSWER D.**

G7C06 What should be the impedance of a low-pass filter as compared to the impedance of the transmission line into which it is inserted?
A. Substantially higher.
B. About the same.
C. Substantially lower.
D. Twice the transmission line impedance.

As we discussed earlier, impedances should always be the same for maximum transfer of power. Therefore, you want the low-pass filter to have the *same impedance* as both the transmission line and the ham transceiver to which it is connected. **ANSWER B.**

G6A05 Which of the following is one effect of lead inductance in a capacitor used at VHF and above?
A. Effective capacitance may be reduced.
B. Voltage rating may be reduced.
C. ESR may be reduced.
D. The polarity of the capacitor might become reversed.

As a Technician Class operator preparing for your General Class ticket, you probably had a chance to look at VHF/UHF equipment circuit boards. Pieces of art, right? These rigs use SMT (surface mount technology) that places components directly on the board to minimize lead inductance. Conventional *capacitors with long lead connection wires have reduced effective capacitance*. **ANSWER A.**

G6A07 Which of the following is a reason not to use wire-wound resistors in an RF circuit?
A. The resistor's tolerance value would not be adequate for such a circuit.
B. The resistor's inductance could make circuit performance unpredictable.
C. The resistor could overheat.
D. The resistor's internal capacitance would detune the circuit.

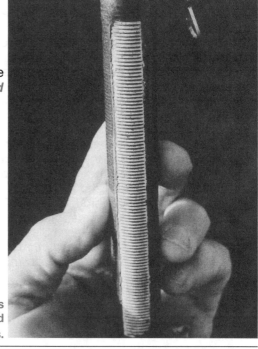

If you were able to scrape off the brown glaze coating on a *wire-wound resistor*, you would quickly see it looks exactly like a coil with evenly spaced turns. Actually, it IS an inductor and would not be suitable in any *circuit* that *could be detuned* accidentally through the use of the wrong component. Normally, we find wire-wound resistors in simple DC applications where we need to drop a small amount of voltage with relatively high current being passed. **ANSWER B.**

A wire-wound resistor like this one is an inductor, and should not be used in tuned circuits.

G6A03 Which of the following is an advantage of ceramic capacitors as compared to other types of capacitors?
A. Tight tolerance.   C. High capacitance for given volume.
B. High stability.   D. Comparatively low cost.

The ceramic capacitor is the workhorse in ham radio equipment. Their reliability is very good, and their comparatively *low cost* helps to keep ham radio equipment reasonably priced. **ANSWER D.**

G6A09 What is an advantage of using a ferrite core toroidal inductor?

A. Large values of inductance may be obtained.
B. The magnetic properties of the core may be optimized for a specific range of frequencies.
C. Most of the magnetic field is contained in the core.
D. All of these choices are correct.

When you pass this test and earn General Class privileges, you may begin to operate High Frequency PORTABLE. A couple of rigs actually have batteries on the inside! With any portable transceiver, including VHF/UHF handhelds, make sure they never get dropped! Dropping radio equipment can fracture the brittle iron cores within a toroidal inductor, which fractures easily. The ferrite core within a toroidal inductor offers large values of inductance, with most of the magnetic field contained within the core so it does not affect other nearby components. The toroidal inductor may be used in applications where core saturation is desirable, so *all of the answer choices* to this question *are correct*. **ANSWER D.**

G6A10 How should the winding axes of solenoid inductors be placed to minimize their mutual inductance?

A. In line.
B. Parallel to each other.
C. At right angles.
D. Interleaved.

Solenoid inductors will interact if they are placed side by side, due to mutual inductance. To minimize this, 2 solenoid inductors should be placed at right angles to their winding axes to minimize unwanted mutual inductance. Just think of how a transformer winding FAVORS mutual inductance, but in this case, *"at right angles"* will MINIMIZE the effect.
ANSWER C.

Inductors are placed at right angles to minimize mutual inductance.

G6A11 Why would it be important to minimize the mutual inductance between two inductors?
 A. To increase the energy transfer between circuits.
 B. To reduce unwanted coupling between circuits.
 C. To reduce conducted emissions.
 D. To increase the self-resonant frequency of the inductors.

If you have ever opened up your radio for a close look at all of the gizmos inside, you probably spotted several copper wire coils near the antenna output jack. Notice that none of these coils are parallel to each other. And the same thing for any coils deep within the receiver or transmitter RF sections of your radio – most coils are at right angles to each other to *eliminate stray coupling* between the RF stages, or are placed in small cans to minimize mutual inductance. **ANSWER B.**

G6A13 What is an effect of inter-turn capacitance in an inductor?
 A. The magnetic field may become inverted.
 B. The inductor may become self resonant at some frequencies.
 C. The permeability will increase.
 D. The voltage rating may be exceeded.

An inductor may be a small coil of wire and, in addition to offering a specific value of inductance, it may inadvertently offer *inter-turn capacitance* at some frequencies. This *could cause the coil to become self-resonant*, because we have now created both inductance and capacitance within a single component. **ANSWER B.**

G5C01 What causes a voltage to appear across the secondary winding of a transformer when an AC voltage source is connected across its primary winding?
 A. Capacitive coupling.   C. Mutual inductance.
 B. Displacement current coupling.   D. Mutual capacitance.

Think of a transformer with interlaced coils. Through mutual inductance within the transformer, voltage applied to the primary will also appear across the secondary. *Mutual inductance.* **ANSWER C.**

G5C02 Which part of a transformer is normally connected to the incoming source of energy?
 A. The secondary.   C. The core.
 B. The primary.   D. The plates.

We normally hook the source of energy to *the primary winding* of a transformer. **ANSWER B.**

G5C06 What is the RMS voltage across a 500-turn secondary winding in a transformer if the 2250-turn primary is connected to 120 VAC?
 A. 2370 volts.   C. 26.7 volts.
 B. 540 volts.   D. 5.9 volts.

This is a turns ratio problem, and is relatively easy to solve using the following equation:

$$E_S = E_P \times \frac{N_S}{N_P} = \frac{E_P \times N_S}{N_P}$$

which means the voltage of the secondary is equal to the voltage of the primary times the number of turns of the secondary divided by the number of turns of the primary. It is derived from the equation that says that the ratio of the secondary voltage, E_S, to the primary voltage, E_P, is equal to the ratio of the turns on the secondary, N_S, to the turns on the primary, N_P.

$$\frac{E_S}{E_P} = \frac{N_S}{N_P}$$

Multiply 120 (E_P) times 500 (N_S), and then divide your answer by 2250. This gives you 26.7 volts. Calculator keystrokes are: Clear, 120 × 500 ÷ 2250 = and the answer is *26.7 volts*. **ANSWER C.**

G5C07 What is the turns ratio of a transformer used to match an audio amplifier having a 600-ohm output impedance to a speaker having a 4-ohm impedance?

    A. 12.2 to 1.   C. 150 to 1.
    B. 24.4 to 1.   D. 300 to 1.

The equation that applies is:

$$\frac{N_P}{N_S} = \sqrt{\frac{Z_P}{Z_S}}$$

The ratio of the turns on the primary, N_P, to the turns on the secondary, N_S, is equal to the square root of the ratio of the primary impedance, Z_P, to the secondary impedance, Z_S. Remember that this turns ratio is primary to secondary.

     Don't worry if you have forgotten about square roots. There's an easy way to solve the problem. The primary impedance, Z_P, of the transformer must match the 600-ohms output impedance of the amplifier; therefore, Z_P is 600 ohms. Divide 600 ohms by 4 ohms, the speaker load impedance on the secondary, and you end up with 150.

     Now you need to find the square root of 150. You know that a square root multiplied by itself gives you the number you want. You can do it by approximation. Since 12 × 12 = 144 and 13 × 13 = 169, you know that the square root of 150 is between 12 and 13. The only answer given that is close is 12.2. Choose it and you have the correct answer. See, you didn't have to remember how to do square roots. The calculator keystrokes are: Clear, *600 ÷ 4 = 150, then press the square root key to produce the answer, 12.25*. **ANSWER A.**

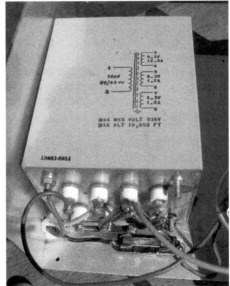

Transformer, from a RADAR.

Good Grounds

CQ FIFTEEN... CQ...CQ FIFTEEN METERS...

G4C05 **What might be the problem if you receive an RF burn when touching your equipment while transmitting on an HF band, assuming the equipment is connected to a ground rod?**

 A. Flat braid rather than round wire has been used for the ground wire.
 B. Insulated wire has been used for the ground wire.
 C. The ground rod is resonant.
 D. The ground wire has high impedance on that frequency.

Nothing worse than petting Fido, while going on the air with your new high frequency installation, and seeing a blue arc from your D-104 microphone to his metal nametag, accompanied by a yelp! Welcome to RF floating along the chassis ground of your equipment. Using wire to ground HF equipment may create an antenna-like circuit that looks like an open circuit to Earth ground. Round ground wires may also behave like wire coils, developing a reactance at specific frequencies, blocking the ground connection to that cold water pipe. This blocking action is called *impedance*. Always ground your equipment with flat braid or copper foil. **ANSWER D.**

G4C07 **What is one good way to avoid unwanted effects of stray RF energy in an amateur station?**

 A. Connect all equipment grounds together.
 B. Install an RF filter in series with the ground wire.
 C. Use a ground loop for best conductivity.
 D. Install a few ferrite beads on the ground wire where it connects to your station.

A safe way to keep all of your equipment at the same ground potential is to run a strip of copper foil at the rear of your station operating desk, and then small, one inch wide copper foil tags going to each radio and accessory. These tags should be long enough to allow you to pull the equipment for servicing, and then accordion-up when you push the gear back into place. *Connecting all of your equipment grounds together* in this way will keep stray RF out of your station. **ANSWER A.**

Good Grounds

G4C09 How can a ground loop be avoided?

A. Connect all ground conductors in series.
B. Connect the AC neutral conductor to the ground wire.
C. Avoid using lock washers and star washers when making ground connections.
D. Connect all ground conductors to a single point.

I have several HF transceivers at the shack, and *all* copper foil *ground connections go to a single ground point*. There, the flattened copper pipe goes through the wall and into the earth. Grounding to a single point will minimize ground loops. If you ever have a new fancy audio stereo system installed in your vehicle, the savvy tech will run all component grounds to a single ground point to minimize ground loops. **ANSWER D.**

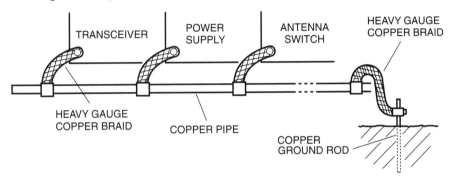

Grounding Equipment

G4C10 What could be a symptom of a ground loop somewhere in your station?

A. You receive reports of "hum" on your station's transmitted signal.
B. The SWR reading for one or more antennas is suddenly very high.
C. An item of station equipment starts to draw excessive amounts of current.
D. You receive reports of harmonic interference from your station.

You are on the air for the first time and everyone reports you sound loud and sort of clear, but with a *hum behind your signal*. Sounds like a ground loop to me, and this can be cured with good copper foil grounding interconnected to each and every piece of metal equipment you have on your desk. This includes that big computer, too. **ANSWER A.**

GROUND FOIL FACTS:
Receive a FREE grounding info package plus a 3″ sample of copper foil by sending an e-mail request to David at Metal-Cable, Inc. His e-mail address is: **david@metal-cable.com**.

Copper foil ground strap provides a good surface area ground.

G4C06 What effect can be caused by a resonant ground connection?
 A. Overheating of ground straps.
 B. Corrosion of the ground rod.
 C. High RF voltages on the enclosures of station equipment.
 D. A ground loop.

If you notice a tingle when you touch anything metal on your equipment during transmit, it is a sign that a *resonant ground connection* is *causing high voltages* to back up *onto the case of the radio*. Time to switch over to some 3-inch wide copper foil. That's what I use here at the "Gordo" station. **ANSWER C.**

G4C01 Which of the following might be useful in reducing RF interference to audio-frequency devices?
 A. Bypass inductor.
 B. Bypass capacitor.
 C. Forward-biased diode
 D. Reverse-biased diode

When you begin operating on General Class frequencies, your powerful, high-frequency SSB transceiver fed into a roof-top antenna system will probably create audio-frequency interference to your own home electronics, and those of your surrounding four neighbors. Bypass capacitors – usually 0.01 mF – will sometimes help minimize this problem when strategically placed across and onto speaker wires and wiring harnesses inside the affected home electronic systems. It's not a cure-all, but *bypass capacitors* are your first step in resolving interference complaint problems on a case-by-case basis. **ANSWER B.**

G4C08 Which of the following would reduce RF interference caused by common-mode current on an audio cable?
 A. Placing a ferrite bead around the cable.
 B. Adding series capacitors to the conductors.
 C. Adding shunt inductors to the conductors.
 D. Adding an additional insulating jacket to the cable.

You can get some terrific audio DSP speakers to add on to older equipment that does not have audio DSP capability. If the speakers are amplified, sometimes RF transmit sounds come out over the speaker itself. You can minimize this common-mode current on the audio cable by *placing ferrite beads around the cable*. Use ferrite beads around ALL the computer and data cables, too. **ANSWER A.**

Snap-on ferrite choke

G4C03 What sound is heard from an audio device or telephone if there is interference from a nearby single-sideband phone transmitter?
A. A steady hum whenever the transmitter is on the air.
B. On-and-off humming or clicking.
C. Distorted speech.
D. Clearly audible speech.

Single-sideband sounds like distorted speech coming over a public address system or certain home electronics. However, double sideband AM CB radio transmissions usually come through loud and clear, so these are easily distinguished from SSB ham transmissions. If someone says you are causing interference, ask the big question: "Does it sound clear, or does it sound garbled?" **ANSWER C.**

G4C04 What is the effect on an audio device or telephone system if there is interference from a nearby CW transmitter?
A. On-and-off humming or clicking.
B. A CW signal at a nearly pure audio frequency.
C. A chirpy CW signal.
D. Severely distorted audio.

CW transmissions come over a PA or home electronics system as *on-and-off humming or clicking sounds*. There is no mistaking the sound of CW. **ANSWER A.**

G4C02 Which of the following could be a cause of interference covering a wide range of frequencies?
A. Not using a balun or line isolator to feed balanced antennas.
B. Lack of rectification of the transmitter's signal in power conductors.
C. Arcing at a poor electrical connection.
D. The use of horizontal rather than vertical antennas.

About once every 6 months I go around and tighten up all connections on my ham equipment. Most important is the tightening-up of the copper ground foil connections to the back of my rig. An *intermittent RF ground* will sometimes create *broadband radio frequency noise* that will magically go away as soon as you give that nut a little clockwise crank. **ANSWER C.**

G4E07 Which of the following is most likely to cause interfering signals to be heard in the receiver of an HF mobile installation in a recent model vehicle?
A. The battery charging system.   C. The anti-theft circuitry.
B. The anti-lock braking system.   D. The vehicle control computer.

The noise blankers on most HF mobile transceivers do a nice job of getting rid of sparkplug pops. But right where you want to operate on the "Gordo" band at 7250 kHz, you hear a steady, raspy carrier that only goes away when you shut off the ignition. Guess what – it may be your *vehicle's onboard computer*. Sometimes these computer noises go all the way up into the 2 meter band as well. Relocating the antenna may help. While I have heard of vehicle computer interference on the VHF and UHF bands, luckily, I almost never hear of this problem down on high frequency. **ANSWER D.**

HF Antennas

G9B10 What is the approximate length for a 1/2-wave dipole antenna cut for 14.250 MHz?

A. 8 feet.   C. 24 feet.
B. 16 feet.   D. 32 feet.

To calculate the length, in feet, of a half-wavelength dipole antenna, divide 468 by the antenna's operating frequency in MHz. Just remember: "2-**4-6-8**, who do we appreciate?" Calculators are permitted in your exam session, so here are the keystrokes to divide 468 by the stated test question frequency: Clear, clear, 468 ÷ 14.250 =. And your answer comes out 32.842 feet, rounded to *32 feet*, Answer D. Yikes! Constructing this dipole, each side will be about 16.4 feet, and how do you go from a decimal point feet to inches? No problem – multiply the decimal feet by 12 and you end up with inches. Remember, the half-wave dipole needs the center insulator to go exactly in the middle. Each side would then be 1/4 wavelength.
ANSWER D.

G9B11 What is the approximate length for a 1/2-wave dipole antenna cut for 3.550 MHz?

A. 42 feet.   C. 131 feet.
B. 84 feet.   D. 263 feet.

Let's do the math: 468 ÷ 3.55 = 131.8 feet! Rounded to *131 feet*. Easy, huh?
ANSWER C.

G9B09 Which of the following is an advantage of a horizontally polarized as compared to vertically polarized HF antenna?

A. Lower ground reflection losses.
B. Lower feed-point impedance.
C. Shorter radials.
D. Lower radiation resistance.

The well-constructed and properly-elevated 1/2 wave dipole will generally outperform ground mounted verticals. This is because the ground mounted vertical usually encounters *ground reflection losses* with everything close by to its base. **ANSWER A.**

G3C11 Which of the following antenna types will be most effective for skip communications on 40 meters during the day?

A. Vertical antennas.
B. Horizontal dipoles placed between 1/8 and 1/4 wavelength above the ground.
C. Left-hand circularly polarized antennas.
D. Right-hand circularly polarized antenna.

A simple *dipole antenna, placed between 1/8 and 1/4 wavelength* above the ground, will give you powerful daytime skywave communications on the 40 meter band. To reach out further, elevate it. To "pull in" your first hop, slightly lower it! **ANSWER B.**

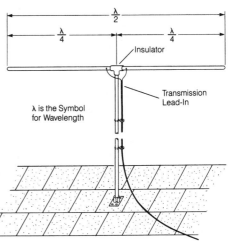

Elmer Point:
The half-wavelength dipole antenna offers no compromise when transmitting and receiving long-range signals. A half-wave dipole usually will outperform multi-band medium length verticals and excel above a fancy $800 mobile, big-coil whip. Dipoles are simple to construct and you can pull up the center section with a rope to configure it as an inverted vee. The only thing that beats a dipole is a big 3-element beam or a quad.

G9B07 How does the feed-point impedance of a 1/2 wave dipole antenna change as the antenna is lowered from 1/4 wave above ground?

A. It steadily increases.
B. It steadily decreases.
C. It peaks at about 1/8 wavelength above ground.
D. It is unaffected by the height above ground.

The 1/2 wave dipole is an outstanding antenna system that will cost you quarters to build. The 1/2 wave dipole, elevated to roof level, will generally outperform expensive, high-frequency mobile whips, expensive trap vertical antennas, and pricey automatic-tuner-fed long wires. But get the 1/2 wave dipole up high. If it is less than 1/4 wavelength above ground, the *feed point impedance will dramatically decrease*, and your radiation pattern goes whacko. **ANSWER B.**

G9B08 How does the feed-point impedance of a 1/2 wave dipole change as the feed-point location is moved from the center toward the ends?
A. It steadily increases.
B. It steadily decreases.
C. It peaks at about 1/8 wavelength from the end.
D. It is unaffected by the location of the feed point.
If you try to offset the feed point of a 1/2 wave dipole from the center toward the ends, *feed point impedance will rise* well above 50 ohms, and your rig's automatic antenna tuner may have a hard time bringing the SWR down to 1:1. **ANSWER A.**

G9B04 What is the low angle azimuthal radiation pattern of an ideal half-wavelength dipole antenna installed 1/2 wavelength high and parallel to the Earth?
A. It is a figure-eight at right angles to the antenna.
B. It is a figure-eight off both ends of the antenna.
C. It is a circle (equal radiation in all directions).
D. It has a pair of lobes on one side of the antenna and a single lobe on the other side.
The dipole gives you a *figure eight pattern at right angles to the antenna* wire. Reception and transmission are minimal off the ends of the wire. **ANSWER A.**

G9B05 How does antenna height affect the horizontal (azimuthal) radiation pattern of a horizontal dipole HF antenna?
A. If the antenna is too high, the pattern becomes unpredictable.
B. Antenna height has no effect on the pattern.
C. If the antenna is less than 1/2 wavelength high, the azimuthal pattern is almost omnidirectional.
D. If the antenna is less than 1/2 wavelength high, radiation off the ends of the wire is eliminated.
Never mount a dipole antenna less than one-half wavelength from the ground for long-range DX. If you do, you may have signal distortion, an *omni-directional radiation pattern*, and most of your signal going straight up. **ANSWER C.**

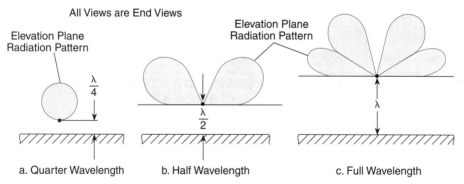

All Views are End Views

Elevation Plane
Radiation Pattern

Elevation Plane
Radiation Pattern

$\frac{\lambda}{4}$

$\frac{\lambda}{2}$

λ

a. Quarter Wavelength   b. Half Wavelength   c. Full Wavelength

The radiation pattern of an antenna changes as height above ground is varied.
Source: *Antennas*, A.J. Evans, K.E. Britain, © 1998, Master Publishing, Niles, Illinois

G9D01 What does the term "NVIS" mean as related to antennas?
A. Nearly Vertical Inductance System.
B. Non-Visible Installation Specification.
C. Non-Varying Impedance Smoothing.
D. Near Vertical Incidence Skywave.

The dipole antenna is fun to play with when strung between two trees using two pulleys. String it way up high, and your angle of radiation lowers, so you get a longer distance skip. However, if you want to talk with your Granddad who is about 300 miles away, try lowering the dipole real close to the ground and notice that the distant skywave skip will fade out and good old Granddad's signal just a few hundred miles away now comes in stronger. This is because your *skywave is nearly vertical*. **ANSWER D.**

G9D03 At what height above ground is an NVIS antenna typically installed?
A. As close to one-half wave as possible.
B. As close to one wavelength as possible.
C. Height is not critical as long as it is significantly more than 1/2 wavelength.
D. Between 1/10 and 1/4 wavelength.

To get your dipole to operate like an NVIS antenna, start near ground level and work from *1/10th to 1/4 wavelength* and listen to close in stations get dramatically stronger. **ANSWER D.**

G9D02 Which of the following is an advantage of an NVIS antenna?
A. Low vertical angle radiation for working stations out to ranges of several thousand kilometers.
B. High vertical angle radiation for working stations within a radius of a few hundred kilometers.
C. High forward gain.
D. All of these choices are correct.

The military and emergency communicators will use NVIS dipoles just a few feet above the ground, causing their *signals* to take off in an almost *vertical direction* to the ionosphere, and coming back down *much closer* in than a regular dipole up one half wavelength. This is great for short skip during the day. **ANSWER B.**

G9B03 What happens to the feed-point impedance of a ground-plane antenna when its radials are changed from horizontal to downward-sloping?
A. It decreases.
B. It increases.
C. It stays the same.
D. It reaches a maximum at an angle of 45 degrees.

The feed point *impedance increases* from 25 ohms to 50 ohms *when you bend the radials downward*. **ANSWER B.**

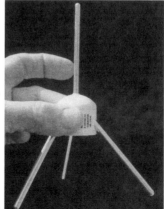

A ground-plane antenna.

G9B02 What is an advantage of downward sloping radials on a quarter wave ground-plane antenna?
  A. They lower the radiation angle.
  B. They bring the feed-point impedance closer to 300 ohms.
  C. They increase the radiation angle.
  D. They bring the feed-point impedance closer to 50 ohms.
On a simple ground-plane antenna, the radials are bent down about 45 degrees to bring the *feed point impedance close to 50 ohms*. If they stick straight out, the impedance is more like 25 ohms, causing a mismatch. **ANSWER D.**

G9B12 What is the approximate length for a 1/4-wave vertical antenna cut for 28.5 MHz?
  A. 8 feet.   C. 16 feet.
  B. 11 feet.   D. 21 feet.
Let's say you want to operate 10 meters mobile. When you operate mobile, the metal frame of the vehicle makes up 1/4 wavelength of your antenna system, and the other 1/4 wavelength is the vertical radiating antenna. Think of your vertical whip as one side of a dipole and the vehicle chassis as the other. First calculate a half wavelength for 28.5 MHz: *468 ÷ 28.5 = 16.42 ÷ 2 = 8.2 feet, rounded to 8 feet*. Sure, you could use "2-3-4" instead of "4-6-8", but I like "4-6-8" because even a mobile antenna is a form of a 1/2 wave dipole. **ANSWER A.**

G9B06 Where should the radial wires of a ground-mounted vertical antenna system be placed?
  A. As high as possible above the ground.
  B. Parallel to the antenna element.
  C. On the surface or buried a few inches below the ground.
  D. At the top of the antenna.
Ground radial wires are important for the ground-mounted vertical antenna to establish its own counterpoise. To hear any real difference when you already have a few radials laid out per band, you must double the number of radials. If you have 2 per band, try 4. If you have 4, try 8. If you need more ground plane than 8 radials per band, try mounting your antenna on the aluminum shed in the backyard. The more ground radials you have on a ground plane antenna, the lower the takeoff angle of radiation. And this means more DX. Get out the shovel, and *start digging to install more ground radials*! **ANSWER C.**

G2D11 Which HF antenna would be the best to use for minimizing interference?
  A. A quarter wave vertical antenna.
  B. An isotropic antenna.
  C. A unidirectional antenna.
  D. An omnidirectional antenna.
Here is a correct answer that is unnecessarily disguised. The term "uni" means single, and the correct answer of a "unidirectional antenna" focusing the signal in one direction to minimize interference in an alternate direction is indeed the correct answer. We normally call a *"unidirectional antenna"* a Yagi, or a beam, or simply directional – but UNIdirectional? **ANSWER C.**

G9B01 What is one disadvantage of a directly fed random-wire antenna?
    A. It must be longer than 1 wavelength.
    B. You may experience RF burns when touching metal objects in your station.
    C. It produces only vertically polarized radiation.
    D. It is not effective on the higher HF bands.

Unless you use a remote-mounted automatic antenna tuner, the random-wire antenna can put a lot of *RF feedback* in your station. **ANSWER B.**

G9C03 Which statement about a three-element, single-band Yagi antenna is true?
    A. The reflector is normally the shortest parasitic element.
    B. The director is normally the shortest parasitic element.
    C. The driven element is the longest parasitic element.
    D. Low feed-point impedance increases bandwidth.

On a three-element beam, the *director is shorter* than the driven element, and the reflector is longer than the driven element. **ANSWER B.**

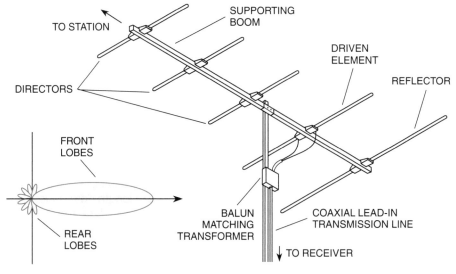

a. Directional Pattern   b. Physical Construction

A beam antenna - the Yagi antenna

Source: *Antennas - Selection and Installation*,©´1986 Master Publishing, Inc., Niles, IL

G9C02 What is the approximate length of the driven element of a Yagi antenna?
    A. 1/4 wavelength.   C. 3/4 wavelength.
    B. 1/2 wavelength.   D. 1 wavelength.

Since the Yagi antenna is a series of dipoles affixed to a boom in the same plane, the *driven element is about one-half wavelength long*. The reflector is a little longer, the directors a little bit shorter. **ANSWER B.**

G9C04 Which statement about a three-element, single-band Yagi antenna is true?
A. The reflector is normally the longest parasitic element.
B. The director is normally the longest parasitic element.
C. The reflector is normally the shortest parasitic element.
D. All of the elements must be the same length.

You can always figure out which way to point a 3 element Yagi by seeing which elements on the boom are the shortest. The *reflector is* normally the *longest* parasitic element, and one or more directors are usually 5% shorter than the driven element. So scope it out, and the shortest elements of the Yagi always point in the general direction of the distant station. Most worldwide Yagis are polarized horizontal, too. **ANSWER A.**

G9C06 Which of the following is a reason why a Yagi antenna is often used for radio communications on the 20 meter band?
A. It provides excellent omnidirectional coverage in the horizontal plane.
B. It is smaller, less expensive and easier to erect than a dipole or vertical antenna.
C. It helps reduce interference from other stations to the side or behind the antenna.
D. It provides the highest possible angle of radiation for the HF bands.

When you pass your General Class license, the Yagi antenna will give you one of the best signals ever on the 20-meter band because it *reduces interference from other stations off to the side or behind*. A small, three-element Yagi works quite nicely just a few feet off the top of your house. **ANSWER C.**

G9C05 How does increasing boom length and adding directors affect a Yagi antenna?
A. Gain increases.   C. Weight decreases.
B. Beamwidth increases.   D. Wind load decreases.

On a worldwide (as well as VHF/UHF) Yagi antenna, boom length determines the amount of gain. The number of elements and the diameter of the elements influence the directivity and bandwidth of the antenna, but the big factor that determines which Yagi antenna is going to outperform another Yagi in *gain* is *boom length*. **ANSWER A.**

A "mobile" CushCraft 5-band, 3-element
Yagi in operation at a Field Day event.

G9C08 What is meant by the "main lobe" of a directive antenna?
A. The magnitude of the maximum vertical angle of radiation.
B. The point of maximum current in a radiating antenna element.
C. The maximum voltage standing wave point on a radiating element.
D. The direction of maximum radiated field strength from the antenna.

The *main lobe* is the *main radiating direction* of the signal. You may use a field strength meter to determine the main lobe of most Yagi antennas. **ANSWER D.**

G9C07 What does "front-to-back ratio" mean in reference to a Yagi antenna?
A. The number of directors versus the number of reflectors.
B. The relative position of the driven element with respect to the reflectors and directors.
C. The power radiated in the major radiation lobe compared to the power radiated in exactly the opposite direction.
D. The ratio of forward gain to dipole gain.

The Yagi antenna gives you an excellent front-to-back ratio. This means that the *majority of the power is radiated out of the front* of the antenna, with *little signal wasted to the back* or to the sides. **ANSWER C.**

G9C09 What is the approximate maximum theoretical forward gain of a three element, single-band Yagi antenna?
A. 9.7 dBi.
B. 9.7 dBd.
C. 5.4 times the gain of a dipole.
D. All of these choices are correct.

This is a great question with a much debated answer – What's the most gain we can get out of a 3 element Yagi? There are plenty of variables including boom length, element construction, feed point techniques… but to get this answer correct, go for answer A, *9.7 dBi* – 9.7 decibels refers to an isotropic antenna which has 0 dBi gain. **ANSWER A.**

G9C10 Which of the following is a Yagi antenna design variable that could be adjusted to optimize forward gain, front-to-back ratio, or SWR bandwidth?
A. The physical length of the boom.
B. The number of elements on the boom.
C. The spacing of each element along the boom.
D. All of these choices are correct.

Any time you begin moving elements along the boom of a Yagi, gain, front to back ratios, and SWR bandwidth will be affected. So, *all of the choices are correct* about the boom length and the number and spacing of the elements along the boom. **ANSWER D.**

G9C01 Which of the following would increase the bandwidth of a Yagi antenna?
A. Larger diameter elements.
B. Closer element spacing.
C. Loading coils in series with the element.
D. Tapered-diameter elements.

The *greater the diameter of the elements*, the *greater the bandwidth* of a Yagi beam antenna for worldwide operation. This is why a wire beam antenna does not offer as much bandwidth as one constructed of large aluminum tubes. **ANSWER A.**

G9C11 What is the purpose of a gamma match used with Yagi antennas?

A. To match the relatively low feed-point impedance to 50 ohms.
B. To match the relatively high feed-point impedance to 50 ohms.
C. To increase the front to back ratio.
D. To increase the main lobe gain.

If you decide to construct your own 3 element beam, or if your Granddad gives you his old 3 element 20 meter monobander, chances are you will need to develop a feed point connection that will take the relatively *LOW feed point impedance up to 50 ohms to match the impedance of your coax*. The gamma match is a series line matching system that lets you tap the driven element at a preferred 50 ohm feed point. To an ohm meter, both the gamma match and the T match may look like a short circuit – but to your radio frequency signal on the 20 meter band for this monoband beam, it will look like a perfect match! **ANSWER A.**

G9C12 Which of the following is an advantage of using a gamma match for impedance matching of a Yagi antenna to 50-ohm coax feed line?

A. It does not require that the elements be insulated from the boom.
B. It does not require any inductors or capacitors.
C. It is useful for matching multiband antennas.
D. All of these choices are correct.

High frequency Yagi antennas may achieve a more predictable radiation pattern by insulating all elements from the boom. Big improvements in those element insulators keep them from cracking when a giant crow lands on one of the elements. Some high frequency antennas *couple the elements directly to the boom through stainless steel hardware and alignment brackets*. This simplifies the matching network to the popular gamma match, where you can hook up the Yagi directly to a 50-ohm coax feed. While the pattern may not be quite as predictable, the gamma match metal-to-boom high frequency antenna is a solid performer. **ANSWER A.**

G9D05 What is the advantage of vertical stacking of horizontally polarized Yagi antennas?

A. Allows quick selection of vertical or horizontal polarization.
B. Allows simultaneous vertical and horizontal polarization.
C. Narrows the main lobe in azimuth.
D. Narrows the main lobe in elevation.

Vertically stacking horizontal Yagis helps *concentrate the main lobe in elevation* for added signal strength to the desired station. Aiming the antenna with your rotator remains about the same. **ANSWER D.**

High frequency beam antennas, on different bands, may be stacked vertically, with little interaction, for a clean look and solid performance. Try for maximum separation to keep the radiation patterns clean.

G9C20 How does the gain of two 3-element horizontally polarized Yagi antennas spaced vertically 1/2 wavelength apart typically compare to the gain of a single 3-element Yagi?

A. Approximately 1.5 dB higher.   C. Approximately 6 dB higher.

B. Approximately 3 dB higher.   D. Approximately 9 dB higher.

You don't need a linear amplifier to boost your effective radiated power output, or increase reception. You could stack a pair of Yagis about one half wavelength apart and pick up a 2 times increase in transmit and receive gain (2x = *3 dB*). **ANSWER B.**

G9D04 What is the primary purpose of antenna traps?

A. To permit multiband operation.

B. To notch spurious frequencies.

C. To provide balanced feed-point impedance.

D. To prevent out of band operation.

A great way to get started on 40 meters, 20 meters, 15 meters, and 10 meters is with a trap dipole or a trap beam antenna. You could also do a trap vertical antenna on the roof. The traps will allow the antenna to naturally tune to specific ham bands. Each trap acts like a stop-band for longer wavelengths, yet offers series loading when you specifically want to operate on longer wavelength lower frequencies. I operate a 4-element 4-band beam antenna and the *traps* work great to give me high frequency operation on *4 specific ham bands*. When you get your new General license, let's try to work each other on the air. **ANSWER A.**

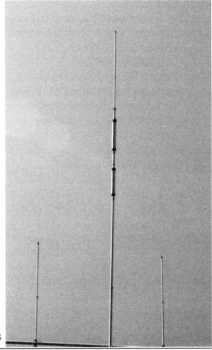

Antenna traps

G9D11 Which of the following is a disadvantage of multiband antennas?

A. They present low impedance on all design frequencies.

B. They must be used with an antenna tuner.

C. They must be fed with open wire line.

D. They have poor harmonic rejection.

Newer ham radio high frequency radios have internal band pass filters to minimize harmonics. A harmonic is a multiple of your fundamental frequency. If you are transmitting on 7 MHz, you would want to minimize any harmonic going out on 14 MHz to nearby hams on that frequency. Unfortunately, while *multiband antennas* have many advantages, they also incur the disadvantage of *poor harmonic rejection*. **ANSWER D.**

G4A06 What type of device is often used to enable matching the transmitter output to an impedance other than 50 ohms?

A. Balanced modulator.   C. Antenna coupler.
B. SWR Bridge.   D. Q Multiplier.

Your new high frequency transceiver is likely all transistorized, including the solid-state RF power amplifier circuit. It is designed to see a perfect 50 ohm match. If the transmitter, with no tuner in line, senses the antenna is not perfect (few are!), the automatic power control circuit will slightly reduce the power output as it senses an elevated SWR. Now you don't want that, do you? The modern transceiver has a built-in antenna coupler. Push the button, and you hear a whir of internal relays, and magically your radio RF power amplifier now sees a 50 ohm match. The built-in antenna "tuner" (coupler) can only match near-resonant antenna systems. They will not tune a long-wire, or a random length dipole. Your match must be close for the antenna tuner to do its job. Time to go out and work on that antenna, rather than mask a problem with the *antenna coupler*! **ANSWER C.**

G9C14 How does the forward gain of a two-element quad antenna compare to the forward gain of a three-element Yagi antenna?

A. About 2/3 as much.   C. About 1.5 times as much.
B. About the same.   D. About twice as much.

If you live in the northeast, severe icing might quickly destroy the 2 element cubicle quad antenna constructed of wire and fiberglass spreaders. If you get a lot of wind, snow, and ice, best consider a 3 element Yagi over the 2 element cubicle quad. *The gain is about the same*. **ANSWER B.**

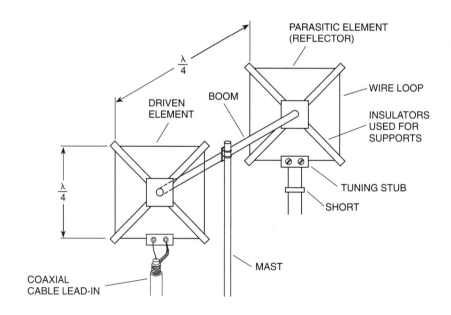

A two-element cubical quad antenna - horizontally polarized
Source: *Antennas—Selection and Installation*, © 1986 Master Publishing, Inc., Niles, IL

G9C13 Approximately how long is each side of a quad antenna driven element?

A. 1/4 wavelength.

B. 1/2 wavelength.

C. 3/4 wavelength.

D. 1 wavelength.

The cubical-quad antenna is a full-wavelength, four-sided, wire antenna system that offers identical to slightly improved performance over a Yagi. *Each side of the cubicle quad is 1/4 wavelength*. This makes the cubicle quad a full wavelength antenna. To determine the length of wire for each side of the driven element, use the following formula:

$$\text{Driven Element for each side (in feet)} = \frac{1005}{f\ (\text{MHz})} \div 4$$

ANSWER A.

G9C19 What configuration of the loops of a two-element quad antenna must be used for the antenna to operate as a beam antenna, assuming one of the elements is used as a reflector?

A. The driven element must be fed with a balun transformer.

B. The driven element must be open-circuited on the side opposite the feed point.

C. The reflector element must be approximately 5% shorter than the driven element.

D. The reflector element must be approximately 5% longer than the driven element.

Remember, *reflectors are generally 5% longer* than the driven element, and director elements are 5% shorter. They ask about the reflector, so it will be 5% longer than the driven element. **ANSWER D.**

G9C15 Approximately how long is each side of a quad antenna reflector element?

A. Slightly less than 1/4 wavelength.

B. Slightly more than 1/4 wavelength.

C. Slightly less than 1/2 wavelength.

D. Slightly more than 1/2 wavelength.

Since the quad antenna is a full wave loop, one side of any of the elements is about 1/4 wavelength. This question asks about the reflector element and, just like in all directional antennas, the *reflector element is slightly longer* than the driven and director elements. **ANSWER B.**

G9C18 What happens when the feed point of a quad antenna is changed from the center of either horizontal wire to the center of either vertical wire?

A. The polarization of the radiated signal changes from horizontal to vertical.

B. The polarization of the radiated signal changes from vertical to horizontal.

C. The direction of the main lobe is reversed.

D. The radiated signal changes to an omnidirectional pattern.

What a work of art – you have built your own cubicle quad, strung between two coconut trees, with each side 1/4 wavelength long for the band of choice. If you feed coax cable to the center of the horizontal side, the polarization will be horizontal to the earth, and this is called... you guessed it... horizontal polarization. If you decide to feed the quad in the center of the *vertical side*, guess what? The radiation will leave the antenna *vertically polarized*. Will all this work of changing from the horizontal to the vertical make any big difference via skywaves? Probably not. **ANSWER A.**

G9C16 How does the gain of a two-element delta-loop beam compare to the gain of a two-element quad antenna?

A. 3 dB higher. C. 2.54 dB higher.
B. 3 dB lower. D. About the same.

The Delta loop beam antenna looks like a triangle with each side 1/3 wavelength long. When compared to the 4 sided cubicle quad antenna, their gain *performance is almost identical*. **ANSWER D.**

G9C17 Approximately how long is each leg of a symmetrical delta-loop antenna?

A. 1/4 wavelength. C. 1/2 wavelength.
B. 1/3 wavelength. D. 2/3 wavelength.

The Delta-loop is another version of a full-wavelength antenna system. The beauty of the delta-loop is you can put the apex of the loop way up at the top of a tree. *Each side* of the symmetrical delta-loop antenna is *1/3 wavelength long*. Use the formula:

$$\text{Driven Element for each side (in feet)} = \frac{1005}{f \text{ (MHz)}} \div 3$$

Just as you did for the quad, divide 1005 by the frequency in MHz. However, since this is a delta-loop, and they only want to know one side, divide your answer by 3 instead of 4. **ANSWER B.**

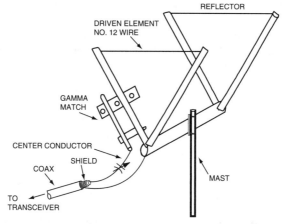

Delta Loop Antenna

G9D07 Which of the following describes a log periodic antenna?

A. Length and spacing of the elements increases logarithmically from one end of the boom to the other.
B. Impedance varies periodically as a function of frequency.
C. Gain varies logarithmically as a function of frequency.
D. SWR varies periodically as a function of boom length.

They call this monster antenna a "logarithmic" periodic antenna because the *length and spacing of the elements increase logarithmically from one end of the boom to the other*. But these are monster antennas with limited applications for specific ham radio bands. Best to go with a multiband dipole or a multiband beam, long before you ever consider the monster "log!" **ANSWER A.**

G9D06 Which of the following is an advantage of a log periodic antenna?
  A. Wide bandwidth.
  B. Higher gain per element than a Yagi antenna.
  C. Harmonic suppression.
  D. Polarization diversity.

Some military stations may use an antenna system called a "log periodic." The elements get progressively shorter on a very long boom, allowing for a *wide bandwidth* over many MHz of radio bands. SWR remains flat on a high frequency log periodic antenna from 10 MHz all the way up to 50 MHz! The only problem with a log antenna is the low forward gain achieved with such a monster in the air. **ANSWER A**

Log periodic antenna

G9D10 Which of the following describes a Beverage antenna?
  A. A vertical antenna constructed from beverage cans.
  B. A broad-band mobile antenna.
  C. A helical antenna for space reception.
  D. A very long and low directional receiving antenna.

If you have plenty of real estate, the *low and long* beverage antenna can achieve some great directionality. This lets you either home-in on a specific signal, or minimize noise coming off a nearby power line. **ANSWER D.**

G9D09 Which of the following is an application for a Beverage antenna?
  A. Directional transmitting for low HF bands.
  B. Directional receiving for low HF bands.
  C. Portable direction finding at higher HF frequencies.
  D. Portable direction finding at lower HF frequencies.

Some of the larger high frequency base station rigs have a separate antenna port for an optional *receive antenna*. This would be perfect for a beverage antenna that you can build yourself out of wire. **ANSWER B.**

G9D08 Why is a Beverage antenna not used for transmitting?
A. Its impedance is too low for effective matching.
B. It has high losses compared to other types of antennas.
C. It has poor directivity.
D. All of these choices are correct.

Sounds like a soft drink, but the beverage is actually a fabulous shortwave receive antenna. It is excellent at pulling in distant stations with minimal atmospheric noise reception. However, use it as a receive only antenna, because its *transmit losses are comparatively higher* than the common dipole. **ANSWER B.**

G4E05 Which of the following most limits the effectiveness of an HF mobile transceiver operating in the 75 meter band?
A. "Picket Fencing" signal variation.
B. The wire gauge of the DC power line to the transceiver.
C. The antenna system.
D. FCC rules limiting mobile output power on the 75 meter band.

75 meters is a great night-time band for talking more than 1,000 miles away. However, to get on 75 meters with a mobile unit you need a monster antenna loading coil and quite possibly a capacity hat on your antenna system. Your HF *antenna system* will be your biggest challenge when operating 75 meters, mobile. **ANSWER C.**

G4E06 What is one disadvantage of using a shortened mobile antenna as opposed to a full size antenna?
A. Short antennas are more likely to cause distortion of transmitted signals.
B. Short antennas can only receive vertically polarized signals.
C. Operating bandwidth may be very limited.
D. Harmonic radiation may increase.

Lightweight, fiberglass/stainless high frequency mobile whips provide some amazing contacts. Mount the whips as high as possible on the vehicle for a good impedance match, and good radiation. Don't mount them down low on a trailer hitch (for safety reasons, too). On 40 meters, the mobile whip will need to be tuned to either the top of the voice band or the bottom of the CW band. On 75 meters, the mobile whip needs to be tuned exactly to where you want to operate, because the *resonant bandwidth may only be 20 or 30 kHz wide*. **ANSWER C.**

G4E02 What is the purpose of a "corona ball" on a HF mobile antenna?
A. To narrow the operating bandwidth of the antenna.
B. To increase the "Q" of the antenna.
C. To reduce the chance of damage if the antenna should strike an object.
D. To reduce high voltage discharge from the tip of the antenna.

When transmitting, the high frequency mobile antenna carries some very high voltage at the tip end of the whip. A big *corona ball* at the tip of the mobile antenna will help *minimize sparks coming off the tip* when you transmit with a mobile linear amplifier in place. Caution: never operate a mobile antenna in proximity to trees. Even at 100 watts, the tip of a mobile antenna, with or without a coronal ball, could arc over to a tree branch near the metal tip. **ANSWER D.**

G4E01 **What is a "capacitance hat" when referring to a mobile antenna?**

A. A device to increase the power handling capacity of a mobile whip antenna.

B. A device that allows automatic band-changing for a mobile antenna.

C. A device to electrically lengthen a physically short antenna.

D. A device that allows remote tuning of a mobile antenna.

What's that eggbeater do-dad on top of my 40 meter high frequency antenna? That's a *capacity hat*, which makes my *antenna look electrically longer* than it physically is. Down on 40, 75, and 160 meters, when operating mobile, you really need a good capacitance hat to gain a little bit more radiation capability. **ANSWER C.**

A Capacity Hat mounted on a 75 meter vertical antenna will improve mobile performance.

G4B11 **Which of the following must be connected to an antenna analyzer when it is being used for SWR measurements?**

A. Receiver.

B. Transmitter.

C. Antenna and feed line.

D. All of these choices are correct.

The portable antenna analyzer is hooked up to the *antenna and feed line* to determine a standing wave ratio (SWR) measurement. **ANSWER C.**

G4B13 **What is a use for an antenna analyzer other than measuring the SWR of an antenna system?**

A. Measuring the front to back ratio of an antenna.

B. Measuring the turns ratio of a power transformer.

C. Determining the impedance of an unknown or unmarked coaxial cable.

D. Determining the gain of a directional antenna.

There is plenty one can do with an antenna analyzer, such as "wringing out" coax cable for opens and shorts, as well as *determining the impedance of that unmarked coax cable* you bought at a local swap meet. **ANSWER C.**

Antenna analyzer

G4B12 What problem can occur when making measurements on an antenna system with an antenna analyzer?
A. SWR readings may be incorrect if the antenna is too close to the Earth.
B. Strong signals from nearby transmitters can affect the accuracy of measurements.
C. The analyzer can be damaged if measurements outside the ham bands are attempted.
D. Connecting the analyzer to an antenna can cause it to absorb harmonics.

This happened to me this past weekend – I was doing some SWR analyzer checks of my 3 element beam after the big rainstorm, and for the first 30 seconds there was a perfect dip right at 14.250 MHz. All of a sudden, my SWR analyzer showed intermittent spikes in the standing wave ratio, looking like some sort of intermittent connection. There were short spikes and long spikes. Hmmmmm, maybe I could even read them as Morse code! It was my neighbor across the street, Jim, N6JF, operating CW on 20 meters with just enough power to cause my SWR analyzer to dance around with dots and dashes. As soon as he stopped transmitting, things settled down to normal. Fact: most SWR analyzers will freeze up and/or become erratic when trying to measure antennas at a repeater site, with strong *signals near the antenna under test*. **ANSWER B.**

G4B08 Which of the following instruments may be used to monitor relative RF output when making antenna and transmitter adjustments?
A. A field-strength meter.   C. A multimeter.
B. An antenna noise bridge.   D. A Q meter.

For relative transmitter output checks, an inexpensive *field-strength meter* is a good addition to any ham station. **ANSWER A.**

Field strength meter

G4B09 Which of the following can be determined with a field strength meter?
A. The radiation resistance of an antenna.
B. The radiation pattern of an antenna.
C. The presence and amount of phase distortion of a transmitter.
D. The presence and amount of amplitude distortion of a transmitter.

A well-calibrated *field-strength meter* might be used to determine *radiation* field *patterns* off that brand-new ground plane antenna you just built for 10 meters. But in the real world of radio, the only way to properly measure antenna pattern field strengths is out in the middle of a 5-acre field where the ground is absolutely flat and there are no buildings or structures to reflect the signal. The antenna is then rotated with the field-strength measurements at a specific fixed spot. Forget about trying to calculate antenna patterns of a vertical antenna in your neighborhood – it just won't work. **ANSWER B.**

G4B07 Which of the following might be a use for a field strength meter?
A. Close-in radio direction-finding.
B. A modulation monitor for a frequency or phase modulation transmitter.
C. An overmodulation indicator for a SSB transmitter.
D. A keying indicator for a RTTY or packet transmitter.

Now here is a "great" correct answer that really describes the best use of a field-strength meter. A fun amateur radio sport is tracking down hidden milliwatt and microwatt transmitters, and the *field-strength meter* will really let you "sniff" an area where you think the *transmitter may be hidden*. T-hunt activities are now a worldwide sport! **ANSWER A.**

Website Resources

▼ IF YOU'RE LOOKING FOR	▼ THEN VISIT
Roll Your Own Antenna	www.cebik.com
Portable Satellite Beams	www.arrowantennas.com
HF Antennas	www.AORUSA.com
	www.directivesystems.com
	www.MFJEnterprises.com
	www.M2INC.com
	www.NEW-Tronics.com
	www.DiamondAntenna.com
	www.Cushcraft.com
	www.Cubex.com
	www.Radioworks.com
	www.Radiowavz.com
	www.alphadelta.com
	www.hiqantennas.com
	www.cq73.com
Coax Cable	www.cableexperts.com
	www.coaxial.com

Coax Cable

G9A02 What are the typical characteristic impedances of coaxial cables used for antenna feed lines at amateur stations?
A. 25 and 30 ohms.   C. 80 and 100 ohms.
B. 50 and 75 ohms.   D. 500 and 750 ohms.

Amateur Radio coax cable usually is rated at *50 ohms* impedance. You might also use some very large TV hard-line coax cable with a proper matching network for your ham setup. Most CATV coax is rated at *75 ohms*. **ANSWER B.**

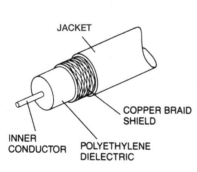

JACKET

COPPER BRAID SHIELD

INNER CONDUCTOR   POLYETHYLENE DIELECTRIC

Coaxial (Called Coax)

This photo shows the different inside construction of various brands of coax cable.

G9A05 How does the attenuation of coaxial cable change as the frequency of the signal it is carrying increases?
A. It is independent of frequency.
B. It increases.
C. It decreases.
D. It reaches a maximum at approximately 18 MHz.

The *higher* you go in *frequency*, the *greater the attenuation* of the transmission line. This is why it's very important to always use the largest size coax cable available for VHF and UHF frequencies. **ANSWER B.**

G9A06 In what values are RF feed line losses usually expressed?
A. Ohms per 1000 ft.
B. dB per 1000 ft.
C. Ohms per 100 ft.
D. dB per 100 ft.

RF feed line losses are usually expressed in *dB (decibels) per 100 feet*. **ANSWER D.**

Attenuation

Frequency (MHz)	(dB/100 ft.)
2	0.21
10	0.5
20	0.71
100	1.7
200	2.4
1000	5.7

Attenuation of RG-8 coax with foam dielectric

G5B10 What percentage of power loss would result from a transmission line loss of 1 dB?
A. 10.9%.
B. 12.2%.
C. 20.5%.
D. 25.9%.

On our examination, it will be expected that certain things are committed to memory within that gray matter. In the real world of ham radio, the memory cells would tell you to look at a chart to calculate decibel losses in a transmission line, or decibel gains in beam antenna effective radiated power. In this question, they ask the percentage loss from a transmission line of 1 dB. One dB of loss results in 0.795 (79.5%) of the energy making it through the coax, leading to 20.5% getting lost in the transmission line (100% - 79.5% = *20.5% loss, the correct answer*). **ANSWER C.**

G9A03 What is the characteristic impedance of flat ribbon TV type twinlead?
A. 50 ohms.
B. 75 ohms.
C. 100 ohms.
D. 300 ohms.

Amateur Radio coax cable usually is rated at 50 ohms impedance. You might also use some very large TV hard-line coax cable with a proper matching network for your ham setup. Most *twin lead is rated at 300 ohms*. **ANSWER D.**

G9A01 Which of the following factors determine the characteristic impedance of a parallel conductor antenna feed line?
A. The distance between the centers of the conductors and the radius of the conductors.
B. The distance between the centers of the conductors and the length of the line.
C. The radius of the conductors and the frequency of the signal.
D. The frequency of the signal and the length of the line.

The characteristic impedance of parallel conductor antenna feed line is most influenced by the *distance between the center of the conductors*, and slightly influenced by actual *radius of the conductors*. An impedance "bump" is caused when the parallel conductor feed line twists and kinks in the wind, dramatically changing the distance between the conductors. Until the feed line gets straightened out, the impedance will no longer be constant. If you are running parallel conductor feed line, keep it absolutely un-kinked, keep it away from metal masts and, if that parallel conductor antenna feed line is part of the vertical radiator, make sure it hangs straight down and never coiled up on itself. **ANSWER A.**

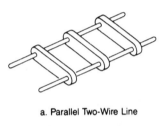

a. Parallel Two-Wire Line

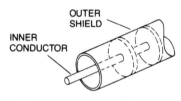

b. Twisted Pair

c. Two-Wire Ribbon Flat Lead (Twin Lead)

INNER CONDUCTOR
OUTER SHIELD

d. Air Coaxial with Washer Insulator

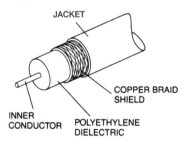

JACKET
COPPER BRAID SHIELD
INNER CONDUCTOR
POLYETHYLENE DIELECTRIC

f. Coaxial (Called Coax)

Different transmission lines
Source: *Antennas – Selection and Installation*, © 1986 Master Publishing, Inc., Niles, IL

G9A07 What must be done to prevent standing waves on an antenna feed line?
- A. The antenna feed point must be at DC ground potential.
- B. The feed line must be cut to an odd number of electrical quarter wavelengths long.
- C. The feed line must be cut to an even number of physical half wavelengths long.
- D. The antenna feed point impedance must be matched to the characteristic impedance of the feed line.

Always try to *match feed point impedance* of the antenna to the characteristic impedance of the feed line for maximum power transfer and minimum SWR. **ANSWER D.**

G9A04 What is the reason for the occurrence of reflected power at the point where a feed line connects to an antenna?
- A. Operating an antenna at its resonant frequency.
- B. Using more transmitter power than the antenna can handle.
- C. A difference between feed line impedance and antenna feed point impedance.
- D. Feeding the antenna with unbalanced feed line.

Keep the impedances of your feed line and antenna the same for minimum standing wave ratio. Standing waves are set-up on the feed line because power is being reflected back due to an *impedance mismatch*. **ANSWER C.**

G4B10 Which of the following can be determined with a directional wattmeter?

A. Standing wave ratio.	C. RF interference.
B. Antenna front-to-back ratio.	D. Radio wave propagation.

A directional watt meter makes a dandy *standing wave ratio checker*. Maximum forward power with minimum reflected power will indicate an acceptable low SWR. **ANSWER A.**

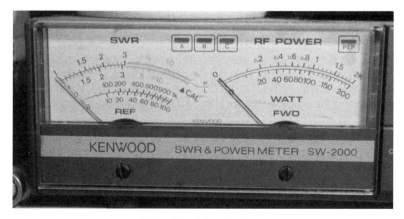

Directional Watt Meter

G9A11 **What standing wave ratio will result from the connection of a 50-ohm feed line to a non-reactive load having a 50-ohm impedance?**

A. 2:1.   C. 50:50.

B. 1:1.   D. 0:0.

50 into 50 is a perfect match, so your *SWR would be 1:1*. **ANSWER B.**

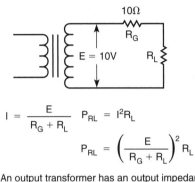

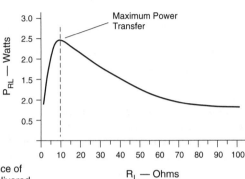

$$I = \frac{E}{R_G + R_L} \qquad P_{RL} = I^2 R_L$$

$$P_{RL} = \left(\frac{E}{R_G + R_L}\right)^2 R_L$$

An output transformer has an output impedance of $R_G = 10$ ohms. We have plotted the power delivered to R_L as R_L varies when $E = 10V$.

The values for the curve are calculated by substituting different values of R_L into the formula when $E = 10V$ and $R_G = 10$.

Maximum power is transferred when $R_G = R_L$.

Maximum power transfer

G9A12 **What would be the SWR if you feed a vertical antenna that has a 25-ohm feed-point impedance with 50-ohm coaxial cable?**

A. 2:1.

B. 2.5:1.

C. 1.25:1.

D. You cannot determine SWR from impedance values.

It's relatively easy to build a 10- or 15-meter ground plane antenna out of copper tubing. The radiating element is one-quarter wavelength long, and the ground radials are also one-quarter wavelength long. The ground radials must be bent down at an approximately 45-degree angle in order to bring the impedance up to 50 ohms, which is the normal impedance of coax cable. If the ground plane copper tubes extend straight out at 90 degrees from the base of the ground plane, the impedance might look like 25 ohms instead of 50 ohms. This would result in an *SWR reading of about 2:1*, acceptable, but easily improved by simply bending the ground radials down at a 45-degree angle. **ANSWER A.**

G9A09 **What standing wave ratio will result from the connection of a 50-ohm feed line to a non-reactive load having a 200-ohm impedance?**

A. 4:1.   C. 2:1.

B. 1:4.   D. 1:2.

50 into 200 goes 4 times, so the impedance of the mismatch is *4:1*. 1:4 is not the correct answer, even though it looks correct. **ANSWER A.**

G9A10 What standing wave ratio will result from the connection of a 50-ohm feed line to a non-reactive load having a 10-ohm impedance?

A. 2:1.   C. 1:5.
B. 50:1.   D. 5:1.

10 into 50 goes 5 times, so the mismatch is *5:1*. 1:5 is not correct. **ANSWER D.**

G9A08 If the SWR on an antenna feed line is 5 to 1, and a matching network at the transmitter end of the feed line is adjusted to 1 to 1 SWR, what is the resulting SWR on the feed line?

A. 1 to 1.
B. 5 to 1.
C. Between 1 to 1 and 5 to 1 depending on the characteristic impedance of the line.
D. Between 1 to 1 and 5 to 1 depending on the reflected power at the transmitter.

Grandpa is on the roof, fixing some shingles when he accidentally chops off a couple of feet of one end of your hidden dipole. The SWR shoots up to 5 to 1, and your new high frequency transceiver immediately pulls back power output. Most new transceivers with built-in automatic antenna tuners can't resolve SWR any higher than 3 to 1, so you drag out an old manual tuner, twiddle the knobs, and magically SWR now drops down to 1:1. Your radio is happy. Grandpa is not. He says every time your are on the air you blank out the TV. Guess what? Even though you reduced the high SWR to a "magic" 1:1 to the transceiver, you still have standing waves flowing back on the outside of the coax from the damaged antenna. This means the feed line is still sky high with an elevated *5 to 1 SWR*. Time to get back to the roof and fix the broken antenna wire. **ANSWER B.**

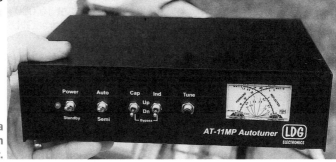

An automatic antenna tuner with built-in SWR meter.

G9A13 What would be the SWR if you feed an antenna that has a 300-ohm feed-point impedance with 50-ohm coaxial cable?

A. 1.5:1.
B. 3:1.
C. 6:1.
D. You cannot determine SWR from impedance values.

The folded dipole has a characteristic impedance of about 300 ohms. This is why it is often fed with twin lead and the impedance transformed with the use of a manual antenna tuner. If you tried to hook up 50-ohm coax cable directly to a 300-ohm feed point, your SWR would be an unacceptable *6:1*. Divide 50 ohms into 300 ohms to calculate the 6:1 ratio. **ANSWER C.**

G6C16 Which of the following describes a type-N connector?

A. A moisture-resistant RF connector useful to 10 GHz.
B. A small bayonet connector used for data circuits.
C. A threaded connector used for hydraulic systems.
D. An audio connector used in surround-sound installations.

As a licensed Technician Class operator, chances are you encountered the Type N connector when working with UHF equipment and microwave equipment. The

Type N connector is great at VHF and UHF antenna feed points because it is *moisture resistant* and it will work all the way up to *10 GHz*, which is 10,000 MHz. When working with your new high frequency SSB transceiver, it is doubtful that you will encounter a Type N connector. **ANSWER A.**

BNC, Type N, and PL 259 Connectors

G6C18 What is a type SMA connector?

A. A large bayonet-type connector usable at power levels in excess of 1 KW.
B. A small threaded connector suitable for signals up to several GHz.
C. A connector designed for serial multiple access signals.
D. A type of push-on connector intended for high-voltage applications.

The modern handheld has switched from an antenna connector called BNC to the *threaded antenna connector called SMA*. Even the microwave operators up at 10,000 MHz use SMA connectors. They are good for several GHz! **ANSWER B.**

SMA connector

G6C13 Which of these connector types is commonly used for RF service at frequencies up to 150 MHz?

A. Octal.   C. PL-259.
B. RJ-11.   D. DB-25.

Your worldwide radio uses an antenna jack called "UHF." This jack will receive the common *PL-259 connector*. On a high frequency radio that might also include 2 meters and 440 MHz, and maybe even 1.2 GHz, you indeed might discover the Type N connector. **ANSWER C.**

Elmer Point: *The common UHF jack on the back of your new HF transceiver will accept the coax cable PL-259 connector, which is the common coax connector for high frequency use. However, the PL-259 is NOT waterproof. If you use an antenna feed point that takes a PL-259, make absolutely sure to completely waterproof the connection with flexible sealant, as well as an added wrap of self-vulcanizing tape. The number one reason why ham radio antenna systems fail is water migrating into the PL-259 and killing the feed point connection. When I visit your QTH, I don't want to see any exposed PL-259 connectors on the coax at the antenna feed point!*

G1B01 What is the maximum height above ground to which an antenna structure may be erected without requiring notification to the FAA and registration with the FCC, provided it is not at or near a public use airport?

A. 50 feet.   C. 200 feet.

B. 100 feet.   D. 300 feet.

The FCC works closely with the FAA when it comes to towers taller than 200 feet. Although you do not need FCC approval for towers less than *200 feet tall*, you may need approval from your city or homeowner's association. [97.15(a)] **ANSWER C.**

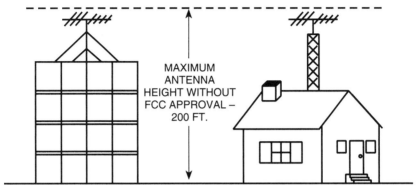

Maximum antenna height

Elmer Point: *When ham radio operators get together for an antenna party, or a club meeting for mobile and base antenna demonstration, one of the greatest safety hazards is the protruding element at eye level. Make absolutely sure all sharp antenna element ends have a soft, protective cover to prevent eye injury. If you are assembling an antenna and ready to put it up on a tower, make certain that everyone is wearing protective safety glasses and a hard hat. Eyesight is so precious and a sharp antenna element at eye level is downright dangerous. Always think eye safety when working around mobile and base antenna systems.*

RF & Electrical Safety

G0A12 What precaution should you take whenever you make adjustments or repairs to an antenna?
- A. Ensure that you and the antenna structure are grounded.
- B. Turn off the transmitter and disconnect the feed line.
- C. Wear a radiation badge.
- D. All of these choices are correct.

If you're going to be working on an antenna system, *turn off the transmitter and disconnect the feed line* from the transmitter so there is absolutely no way that someone could accidentally transmit while you are aloft repairing the antenna. **ANSWER B.**

G0B08 What should be done by any person preparing to climb a tower that supports electrically powered devices?
- A. Notify the electric company that a person will be working on the tower.
- B. Make sure all circuits that supply power to the tower are locked out and tagged.
- C. Unground the base of the tower.
- D. All of these choices are correct .

To insure that no one starts rotating a beam antenna using the electric powered rotator on the top of the tower, double check that the rotator control unit down below is *unplugged and tagged "climber aloft."* **ANSWER B.**

G0B07 Which of the following should be observed for safety when climbing on a tower using a safety belt or harness?
- A. Never lean back and rely on the belt alone to support your weight.
- B. Always attach the belt safety hook to the belt D-ring with the hook opening away from the tower.
- C. Ensure that all heavy tools are securely fastened to the belt D-ring.
- D. Make sure that your belt is grounded at all times.

Fall prevention from a tower is best achieved with a professional safety harness that regularly gets inspected, making sure the belt's *safety hook attaches to the "D" ring with the hook always away from the tower.* **ANSWER B.**

Before going aloft to repair an antenna, always disconnect the feedline from the transmitter to prevent accidental exposure to RF radiation.

G0A08 Which of the following steps must an amateur operator take to ensure compliance with RF safety regulations when transmitter power exceeds levels specified in part 97.13?
    A. Post a copy of FCC Part 97 in the station.
    B. Post a copy of OET Bulletin 65 in the station.
    C. Perform a routine RF exposure evaluation.
    D. All of these choices are correct.
Performing a *routine RF exposure evaluation* is a good idea for all amateurs to ensure compliance with RF-safety regulations, and to ensure that you and your neighbors are not becoming "overexposed" to RF radiation. **ANSWER C.**

G0A11 What precaution should you take if you install an indoor transmitting antenna?
    A. Locate the antenna close to your operating position to minimize feed-line radiation.
    B. Position the antenna along the edge of a wall to reduce parasitic radiation.
    C. Make sure that MPE limits are not exceeded in occupied areas.
    D. No special precautions are necessary if SSB and CW are the only modes used.
It is always a good idea to locate an antenna as far away as possible from living spaces that will be occupied when you are transmitting on the air. If you're just receiving, no problem – but when you're transmitting, to minimize RF exposure get your wire antenna away from everyone! This will help assure that Maximum Permissible Exposure (MPE) *limits are not exceeded.* **ANSWER C.**

G0A01 What is one way that RF energy can affect human body tissue?
    A. It heats body tissue.
    B. It causes radiation poisoning.
    C. It causes the blood count to reach a dangerously low level.
    D. It cools body tissue.

When you reheat a slice of ham or beef in the microwave oven, what are you actually doing? The microwave oven concentrates radio signals into the meat and it heats up. If microwave RF energy from your antenna system is concentrated on the human body, it *heats the body tissue* just like your leftovers. **ANSWER A.**

Never stand in front of a microwave feedhorn antenna.
On transmit, it radiates a concentrated beam of RF energy.

G0A03 How can you determine that your station complies with FCC RF exposure regulations?
 A. By calculation based on FCC OET Bulletin 65.
 B. By calculation based on computer modeling.
 C. By measurement of field strength using calibrated equipment.
 D. All of these choices are correct.

You can determine how your station complies with FCC RF exposure regulations by using The W5YI RF Safety Tables in the Appendix of this book. You can use the tables to estimate safe distances based on FCC OET Bulletin No. 65, or by your own calculations based on computer modeling. You also can actually go out there and measure the power density levels with a field strength meter making sure to use calibrated equipment. *All of these choices* are a good way to determine whether or not you are going to expose yourself or your neighbors unnecessarily. [97.13(c)(1)] **ANSWER D.**

G0A04 What does "time averaging" mean in reference to RF radiation exposure?
 A. The average time of day when the exposure occurs.
 B. The average time it takes RF radiation to have any long-term effect on the body.
 C. The total time of the exposure.
 D. The total RF exposure averaged over a certain time.

Time-averaging is a method of *calculating* an individual's *total exposure to RF radiation over a given period of time*. The premise of time-averaging is that the human body can tolerate larger amounts of RF radiation if the exposure is received in short "bursts" as compared to a constant exposure at the same high level. As depicted here, total exposure to various levels of radiation is averaged over a 6-minute period. On a time-averaged basis, the amount of thermal load on the body is equal in all cases. **ANSWER D.**

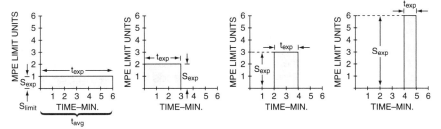

The general equation for time averaging exposure equivalence is: $S_{exp} \, t_{exp} = S_{limit} \, t_{avg}$

G0A07 What effect does transmitter duty cycle have when evaluating RF exposure?
A. A lower transmitter duty cycle permits greater short-term exposure levels.
B. A higher transmitter duty cycle permits greater short-term exposure levels.
C. Low duty cycle transmitters are exempt from RF exposure evaluation requirements.
D. High duty cycle transmitters are exempt from RF exposure requirements.

Duty cycle is the percentage of time the transmitter is actually sending out energy. If you hold down your telegraph key for continuous full-transmit-power output, this would be 100% duty cycle. If the space in between each dit were equal to the duration of the transmitted signal, this would be a 50% duty cycle. The *lower duty cycle will permit greater short-term exposure levels to RF radiation*. **ANSWER A.**

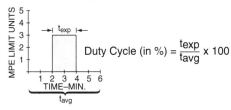

$$\text{Duty Cycle (in \%)} = \frac{t_{exp}}{t_{avg}} \times 100$$

Duty Cycle

G0A02 Which of the following properties is important in estimating whether an RF signal exceeds the maximum permissible exposure (MPE)?
A. Its duty cycle.   C. Its power density.
B. Its frequency.   D. All of these choices are correct.

As a new General Class operator, you do not want to expose yourself, your family and your neighbors to excessive RF power outputs. Any time you put up a new radio system do the following:
1. Compute duty cycle – if you're doing digital, your duty cycle will be much higher than voice or Morse code.
2. Compute the frequency – higher bands may require lower power outputs.
3. Compute power density – How close are you and your family, and your neighbors, to that radiating antenna?

Do all of these steps. There is a lot to do before considering your station SAFE and not exceeding the maximum permissible exposure (MPE) limits. **ANSWER D.**

G0A05 What must you do if an evaluation of your station shows RF energy radiated from your station exceeds permissible limits?
> A. Take action to prevent human exposure to the excessive RF fields.
> B. File an Environmental Impact Statement (EIS-97) with the FCC.
> C. Secure written permission from your neighbors to operate above the controlled MPE limits.
> D. All of these choices are correct.

Federal Communications Commission rules state that: "The licensee must perform the routine RF environmental evaluation prescribed by Section 1.1307(b) of this chapter, if the power of the licensee's station exceeds the following PEP limits." *If your station exceeds permissible limits, you need to make changes to solve the problem*. **ANSWER A.**

G0A09 What type of instrument can be used to accurately measure an RF field?
> A. A receiver with an S meter.
> B. A calibrated field-strength meter with a calibrated antenna.
> C. A betascope with a dummy antenna calibrated at 50 ohms.
> D. An oscilloscope with a high-stability crystal marker generator.

Although a simple field-strength meter can show the presence of radio frequency emissions, it takes a precisely *calibrated field-strength meter with a calibrated antenna* to accurately measure RF fields. **ANSWER B.**

G0A13 What precaution should be taken when installing a ground-mounted antenna?
> A. It should not be installed higher than you can reach.
> B. It should not be installed in a wet area.
> C. It should limited to 10 feet in height.
> D. It should be installed so no one can be exposed to RF radiation in excess of maximum permissible limits.

There are all sorts of great 5-band and 7-band trap vertical antennas that work nicely on the ground, or on a metal shed just above the ground. When you're looking for a spot to mount the antenna, keep in mind maximum permissible exposure limits and *install it so that no one can actually walk up to it*, or stand near it, when you are transmitting. This goes for your pets, too. Don't fry Fido or Furball by leaving a ground-mounted vertical unfenced. **ANSWER D.**

Ground-mounted antennas, like this 10GHz EME dish in Alaska, should be surrounded by a wood safety fence. And make sure to stay a safe distance away from any antenna on transmit!

G0A10 What is one thing that can be done if evaluation shows that a neighbor might receive more than the allowable limit of RF exposure from the main lobe of a directional antenna?
- A. Change from horizontal polarization to vertical polarization.
- B. Change from horizontal polarization to circular polarization.
- C. Use an antenna with a higher front-to-back ratio.
- D. Take precautions to ensure that the antenna cannot be pointed in their direction.

If your calculations indicate you may be exposing your neighbors to too much RF, you may need to relocate your entire antenna system, or take precautions to *ensure it cannot be pointed at their house* when transmitting. **ANSWER D.**

G0B14 Which of the following is covered by the National Electrical Code?
- A. Acceptable bandwidth limits.
- B. Acceptable modulation limits.
- C. Electrical safety inside the ham shack.
- D. RF exposure limits of the human body.

The *National Electrical Code* covers *electrical safety standards* as they relate to conductors and wiring inside your ham shack. RF exposure limits to the human body are covered by ANSI, not by the NEC. **ANSWER C.**

G0B05 Which of the following conditions will cause a Ground Fault Circuit Interrupter (GFCI) to disconnect the 120 or 240 Volt AC line power to a device?
- A. Current flowing from one or more of the hot wires to the neutral wire.
- B. Current flowing from one or more of the hot wires directly to ground.
- C. Over-voltage on the hot wire.
- D. All of these choices are correct.

Ground fault circuit interrupters are found in newer electrical sockets, and they instantly open a circuit when they detect *current flowing from the hot wire to ground*. **ANSWER B.**

Ground fault receptacle

G0B06 Why must the metal enclosure of every item of station equipment be grounded?
- A. It prevents blowing of fuses in case of an internal short circuit.
- B. It prevents signal overload.
- C. It ensures that the neutral wire is grounded.
- D. It ensures that hazardous voltages cannot appear on the chassis.

Good station grounding insures that *no hazardous* or dangerous *voltages* appear *on the metal chassis* of your equipment. **ANSWER D.**

G0B01 Which wire or wires in a four-conductor line cord should be attached to fuses or circuit breakers in a device operated from a 240-VAC single-phase source?

A. Only the hot wires.
B. Only the neutral wire.
C. Only the ground wire.
D. All wires.

Most ham transceivers run straight out of the box on 12 volts DC. The big huge honking high frequency base stations may have their own power supplies built-in – 110 VAC. However, kilowatt linear amplifiers may require 240 volts alternating current, and for single phase *240 VAC we fuse both the hot black and red wires*, and NEVER fuse the neutral white wire and NEVER fuse the neutral ground bare copper wire. Most amplifiers already have the fuse circuits built in. **ANSWER A.**

G0B03 Which size of fuse or circuit breaker would be appropriate to use with a circuit that uses AWG number 14 wiring?

A. 100 amperes.
B. 60 amperes.
C. 30 amperes.
D. 15 amperes.

This one works out "F" for "Fourteen" (14 gauge) wiring that will handle *Fifteen (15) amps*. **ANSWER D.**

Wire Size A.W.G. (B&S)	Current-Amps (Continuous Duty)	
	Single Wire	Bundled Wire
8	73	46
10	55	33
12	41	23
14	32	17
16	22	13
18	16	10

American Wire Gauge (AWG) wire size vs. current capability

G0B02 What is the minimum wire size that may be safely used for a circuit that draws up to 20 amperes of continuous current?

A. AWG number 20.
B. AWG number 16.
C. AWG number 12.
D. AWG number 8.

The minimum wire size to handle 20 amps should be AWG #12. Remember *"T" for "Twelve" (12) and "T" for "Twenty" (20)*. **ANSWER C.**

G0B12 What is the purpose of a transmitter power supply interlock?

A. To prevent unauthorized access to a transmitter.
B. To guarantee that you cannot accidentally transmit out of band.
C. To ensure that dangerous voltages are removed if the cabinet is opened.
D. To shut off the transmitter if too much current is drawn .

Linear amplifiers using vacuum tubes will likely have a power supply interlock switch. If the equipment is on and *the power supply door is opened*, the equipment will either *instantly power off*, or you will be greeted by a loud "bang" when the high voltage gets shorted to ground tripping an internal fuse or breaker. **ANSWER C.**

G0B10 Which of the following is a danger from lead-tin solder?

A. Lead can contaminate food if hands are not washed carefully after handling.
B. High voltages can cause lead-tin solder to disintegrate suddenly.
C. Tin in the solder can "cold flow" causing shorts in the circuit.
D. RF energy can convert the lead into a poisonous gas.

Any time you are working on radio equipment innards, after you get through with your great soldering job always *wash your hands* to remove any lead that might be contained in the roll of Granddad's old solder. **ANSWER A.**

G0B11 Which of the following is good engineering practice for lightning protection grounds?

A. They must be bonded to all buried water and gas lines.
B. Bends in ground wires must be made as close as possible to a right angle.
C. Lightning grounds must be connected to all ungrounded wiring.
D. They must be bonded together with all other grounds.

Good engineering practice for a lightning ground indicates that *all grounds be bonded together* with all other grounds. **ANSWER D.**

G0B09 Why should soldered joints not be used with the wires that connect the base of a tower to a system of ground rods?

A. The resistance of solder is too high.
B. Solder flux will prevent a low conductivity connection.
C. Solder has too high a dielectric constant to provide adequate lightning protection.
D. A soldered joint will likely be destroyed by the heat of a lightning strike.

Carefully inspect tower grounding circuits where all ground connections have been swaged rather than soldered. *Solder might melt on a direct lightning strike*. **ANSWER D.**

G0B04 Which of the following is a primary reason for not placing a gasoline-fueled generator inside an occupied area?

A. Danger of carbon monoxide poisoning.
B. Danger of engine over torque.
C. Lack of oxygen for adequate combustion.
D. Lack of nitrogen for adequate combustion.

If you plan to operate from a ham radio emergency communications vehicle, make sure you spot the carbon monoxide alarm system. In rare cases, the running generator beneath the communications vehicle might leak *carbon monoxide* into the operating compartment. This could become LETHAL. Always think "fresh air" safety when operating in a vehicle with the engine or generator running. **ANSWER A.**

G0B13 What must you do when powering your house from an emergency generator?

A. Disconnect the incoming utility power feed.
B. Insure that the generator is not grounded.
C. Insure that all lightning grounds are disconnected.
D. All of these choices are correct.

In an emergency when you look around and see nothing but downed power poles and know it is going to be several days – or weeks – before electricity is restored, you may wish power your house with your RV generator. Any time you back-feed voltage into your home, you MUST ABSOLUTELY trip all of the breakers to *disconnect the incoming utility power feed* to make sure you are not sending voltage beyond your electrical panel. This could prove dangerous for anyone working on supposedly dead power lines. It will also keep your RV generator from being blown up when they unexpectedly reconnect you to the grid. Disconnect the incoming utility power feed at the panel! **ANSWER A.**

G0B15 Which of the following is true of an emergency generator installation?
    A. The generator should be located in a well ventilated area.
    B. The generator should be insulated from ground.
    C. Fuel should be stored near the generator for rapid refueling in case of an
       emergency.
    D. All of these choices are correct.

When you set up for field day in June, chances are you'll be using an emergency generator to keep all of the stations on the air. Make sure the generator is *located in a well-ventilated area*; not where some of the field day operators will inhale the toxic exhaust. Make sure the generator is properly grounded, and always store the generator's fuel in a safe place away from any inhabited area. **ANSWER A.**

4

Taking the General Class Examination

Get ready for worldwide skywave band privileges! As soon as you pass your General Class theory exam, you will be licensed to broadcast on frequency bands that regularly offer skywave excitement every hour of the day, and all through the night. This chapter tells you how the examination will be given, who is qualified to administer the General Class exam, and what happens after you successfully complete the written exam.

IMPORTANT NOTE: In order to receive credit for Technician Class, a prerequisite to the Element 3 General Class examination, you must bring a photocopy of your original, signed, valid Technician Class license to the examination site. You also will need 2 forms of identification, one of which must be a photo ID (such as your driver's license). If you can't find your Technician Class original license, you can download a copy from the FCC website at **wireless.fcc.gov** and then doing a license search by your call sign or your name. There no longer is any requirement to show a Morse code credit or to take a Morse code test.

If you are brand new to ham radio, you can certainly take both the Technician and General Class exams in one test session as long as you pass the Technician exam first.

THE GENERAL CLASS EXAMINATION

Here is an overview of the General Class examination and what to expect when you go to the test session.

Examination Administration

The General Class exam is given by a team of 3 Volunteer Examiners (VEs) – hams who hold Advanced or Extra Class licenses and who are accredited to administer your exam by a Volunteer Examiner Coordinator (VEC).

Volunteer Examiner Teams offer examinations on a regular basis at local sites to serve their communities. Generally, the VECs closely coordinate their activities with one another, so you should be able to find a nearby test site and exam date that is convenient for you. You can obtain information about VECs and exam sessions in your area by checking with your local radio club, ham radio store, or local packet bulletin boards. A list of VECs that was current at the time of publication is given in the Appendix (see page 203).

Once you have found your local VE team, contact them to select a test date and location and pre-register for your examination. They will hold a seat for you at the next available session. Don't be a no-show, and don't be a surprise-show. Call them ahead of time and pre-register!

The Volunteer Examiners are not compensated for their time and skills, but they are permitted to charge you a fee for certain reimbursable expenses incurred in preparing, administering, and processing the examination. The maximum fee is adjusted frequently and currently is about $14.00. When you call to make your exam reservation, ask the VE the current amount of the exam fee.

Want to find a test site fast?
Visit the W5YI-VEC website at: www.w5yi.org, or call them at 800-669-9594.

EXAM CONTENT

The questions, answers, and distracters for each question of the General Class written examination are public information. The question pool included in this book contains all 456 possible questions that can be used to make up your 35-question Element 3 written examination. The VEC is not permitted to change any of the wording, punctuation or numerical values included in any questions, answers, or distracters. The VEC can change the A-B-C-D order of the answers, if it wishes.

Also, the VEC is required to select one question from each syllabus topic under each subelement so you should expect an exam that contains one question from each syllabus topic within each subelement. Look again at the question pool syllabus on page 209 to see how the test will be constructed.

WHAT TO BRING TO THE SITE

Here's what you'll need to bring with you for your General Class examination:
- The fee of approximately $14.00.
- The original plus two copies of your current Technician or Technician-Plus license. If your license has not yet arrived, make sure you bring the original plus two copies of your Certificate of Successful Completion of Examination (CSCE) indicating your most current license status.
- A photo identification card – your driver's license is ideal for this purpose.
- Some sharp pencils and fine-tip pens. It's good to have a backup.
- Calculators may be used. However, the examiners may erase the memory before your exam begins.
- Any other items that the VEC asks you to bring.

TAKING THE EXAM

Don't speed read the examination! Read each question carefully. Take your time looking for the correct answer. Some answers start out looking correct, but end up wrong. When you finish, go back over every question and double-check your answers. When you are satisfied that you have passed the examination, turn in all of your test papers to the examination team. Make sure to thank your VEs and to let them know how much you appreciate their efforts to help promote our hobby.

AFTER THE EXAM

Wait patiently outside the exam room for your results. Chances are the VEs will greet you with a smile and your CSCE. If you didn't pass, they will tell you what to do next. When you are told you passed the exam, be sure you are given the appropriate paperwork:

- The CSCE, signed by all three examiners.
- Make sure the temporary identifier is filled in so you can use it to immediately go on the air with your new General Class frequency privileges.

COMPLETING NCVEC FORM 605

When you arrive at the examination site, one of the first things you will do is complete the NCVEC Form 605. This form is retained by the Volunteer Exam Coordinator who transfers your printed information to an electronic file and sends it to the FCC for your new license, or upgrade. Your application may be delayed or kicked-back to you if the VEC can't read your writing. Make absolutely sure you print as legibly as you can, and carefully follow the instructions on the form.

NCVEC Form 605

Name

If you are upgrading from a current license, it is very important to compare the information on your present license with what you are writing on NCVEC Form 605. Make sure that *everything* on NCVEC Form 605 *is identical* to how your present license reads. Fill in your last name, first name, middle initial, and suffix such as junior or senior. You must stay absolutely consistent with your name on any future Form 605s for upgrades or changes of address. If you start out as "Jack" and end up "John," the computer will throw out your next application. If you decide to use a nickname, this is okay – but down the line when you visit a foreign country, they may ask you for identification that needs to illustrate this same nickname. It is best to stick with the name that is on most of your personal pictured IDs, such as your Driver's License.

Date of Birth

The biggest problem here is, without thinking, you put down this year's date rather than the year in which you were born. One out of twenty make this mistake.

Social Security Number & FRN

You are required to write in either your Social Security Number or your FCC Registration Number (FRN) in the designated box. If you are a citizen of another country, put down the country name in this box. Your current Technician Class license should show your FRN. If you do not have an FRN and you prefer not to disclose your Social Security Number, you should obtain an FRN *before* the exam session so that you can complete your Form 605 when you pass your test. To obtain an FCC Registration Number, go to the following website and follow the instructions there. Visit:

https://fjallfoss.fcc.gov/coresWeb/publicHome.do

Take our word for it – just give them your SSN and avoid a lot of grief.

Address

Have you moved? Check your current Technician Class license. Did you write in the same EXACT address? If so, you are good to go. If your address is DIFFERENT, check the "change" box and list your new address.

e-mail Address

This is optional, but it's a good idea because the amateur radio service is now under the FCC's Universal Licensing System. Once you get your new call sign, you will be able to work with the FCC directly via computer, including change of address, change of name, and license renewals without having to do any paperwork.

Phone Numbers

There are two boxes for phone numbers – one for a daytime contact, and the other for your FAX number. Put both numbers down in just in case the VEC or VE team need to re-contact you because they can't read your writing.

Signature

Sign your name as legibly as possible and include all of the letters that you printed as your name at the top of the form. Don't just put down a squiggle or an initial. You need to sign your name all the way out, including all of the letters that were in your printed name.

Final Check

Finally, double-check that your handwriting is legible. If a single letter in your name can't be read clearly and is misinterpreted, subsequent electronic filings may get returned as no action. Make sure your Form 605 is as clear as a bell to your Volunteer Examination team, who will then forward it to their VEC.

Your Examiners' Portion

The VEC will carefully review your NCVEC Form 605 to ensure that they can read your handwriting and that everything looks okay. They will then enter this information into their computer database, and will most likely file your test passing results electronically to their Volunteer Examiner Coordinator. The VEC will then verify the information and electronically file your results with the FCC.

YOUR UPGRADE TO GENERAL CLASS

Usually, your General Class upgrade or new call sign will be granted within 72 hours of passing an examination, if it is electronically filed by the VEC – and you should receive a paper copy of your new General Class license in about 3 weeks. When you pass your written exam, you will be issued a CSCE – Certificate of Successful Completion of Examination. This CSCE allows you to begin using your new privileges *immediately*. After your call sign, append the letters "AG" to indicate your upgrade is being processed by the FCC. As soon as you see your upgrade on the electronic database, or receive your new license, you can drop the "AG" at the end of your call sign. Visit the W5YI-VEC website, where you will find links to other sites that allow you to look-up your upgrade or new call sign. The address is **www.w5yi.org**.

If you are going from no license to General Class, *you may not operate immediately* with your CSCE because you have no call sign. But thanks to electronic filing, your new call sign should show up on the FCC database in about 3 to 5 days.

GENERAL CLASS CALL SIGNS

The FCC has exhausted the availability of "Group C" Technician and General Class call signs that begin with the letter "N," a number, and three other letters, such as N9ABC.

U.S. Call Sign Areas

If you did not check the "Change Call Sign" box on your NCVEC Form 605 application, you will simply keep your current call sign. However, if you do check the "Change Call Sign" box, you will receive an entry-level "Group D" call sign as if you were a newly-licensed amateur. I suggest you don't check the "Change Call Sign" box and stick with your present call sign.

Vanity Call Signs

You are eligible to replace your computer-generated, no-choice call sign with a vanity call sign of your choosing. This call sign could be made up of your initials, or represent your love of animals (K9DOG) or could be call letters that your late mom or dad had when they got started in ham radio years ago. General Class amateur operators may request a vanity call sign from Group D or Group C. You also may request a call sign that was previously assigned to you that may have expired years ago, as well as a call sign of a close relative or former holder who is now deceased.

There is an additional fee charged by the FCC for vanity call signs and also an additional fee for renewal of a vanity call sign. But thousands of amateur enjoy a call sign that has special meaning to them personally, or just sounds better phonetically. You can file for a vanity call direct with the FCC by mail using the required application Form 605 and remittance form159, or electronically on the FCC web site, if you know your FRN and password to access your license record. Today, the best way to file a vanity application is to file electronically. To make it easier for you and to insure you get the vanity call sign that you want, the W5YI Group at 800-669-9594 offers a vanity filing service and will electronically file your application for you. For an nominal service fee, they will research when and if a specific call is available, file your application and pay the FCC filing fee for you. You could end up with the exact call sign of your choice using their 99% success rate vanity filing service. You can enter your application and call sign choices right on their web site at www.w5yi.org.

As for me, I'm staying with my original-issue WB6NOA call sign. If I changed it, I would be breaking a 50-year tradition!

CONGRATULATIONS! YOU PASSED!

After you pass the written exam and code test, congratulations are in order and we offer you a big welcome to the worldwide privileges of General Class! Day or night, summer or fall, sunshine or rain, there is always a worldwide band open and ready for General Class voice, code, PACTOR, and television communications. Your new worldwide privileges are added to your existing VHF and UHF privileges. Before you go on the air, do a lot of listening. This will assure that you get started on the right foot with your new privileges. Remember, even the worldwide bands have band plans, so make sure you are operating within the plan for your worldwide communications.

I also would like you to write me so I can send you an exclusive General Class passing certificate, plus some valuable manufacturers' discount coupons. Send a self-addressed large envelope with 12 first class stamps loose on the inside to: Gordon West, WB6NOA, Radio School, Inc., 2414 College Drive, Costa Mesa, CA 92626.

Once again... *Welcome to General Class!* I hope to work you on one of the worldwide bands very soon. You can catch me mobile in our communications van regularly on 14.240 MHz, and I'm a "regular" on 10 meters from my Southern California home at 28.400 MHz.

Your next upgrade is *Extra Class.* For Extra Class, use my 3rd Ham Book, *Extra Class.* My explanations make the formula problem-solving easy and the learning fun! So start thinking about that upgrade now.

It's been fun teaching you the General Class. Good luck on that upcoming exam – I know you're going to pass!

Gordon West, WB6NOA

PS: Keep reading! Chapter 5 will get you started on your journey to learn CW!
Aw, come on – you can do it!!

5

Learning Morse Code

On April 15, 2000, the FCC dropped the 20- and 13-word-per-minute Morse code requirements for worldwide frequency privileges for General and Extra Class operators down to 5 words per minute. On February 23, 2007, the FCC *totally eliminated* the Morse code test as a prerequisite for high frequency operation.

The elimination of the Morse code test for operation on worldwide frequencies conforms to international radio regulations. In 2003, the International Telecommunications Union (ITU) World Radiocommunication Conference voted to allow individual nations to determine whether or not to retain a Morse code test as a requirement to operate on frequencies below 30 MHz.

When the FCC eliminated the Morse code test for Technician Class operators in 1991 for VHF/UHF operating, the ruling was adopted with little opposition. However, the announcement that the FCC was considering total elimination of the Morse code test drew thousands of written comments to the FCC. Many comments supported code test elimination, while a minority urged the FCC to retain a code test because of the strong tradition of CW as a ham radio operating mode.

The Federal Communications Commission concluded that "…this change (eliminating the code test) eliminates an unnecessary burden that may discourage current amateur radio operators from advancing their skills and participating more fully in the benefits of amateur radio." The FCC Commissioners recognized the simple fact that learning Morse code was keeping many very technical, talented hams from obtaining their General and Extra Class licenses. Morse code is much like musical rhythms. Some people are tone deaf, and some people couldn't carry a rhythm in a hand basket.

So for years, the Morse code test was an insurmountable hurdle to many talented Technician Class hams who wanted to upgrade. If I could take these Techs and put them into one of my regular Morse code classes, we usually could get the majority of them through the CW test with outside home study, on the air practice (on 2 meters), and classroom study followed by the code test. But throughout the country, Morse code classes were few and far between and it is tough to learn new music and a new language without classroom instruction.

Technician Class operators could not practice on the worldwide airwaves to learn the code because these bands were reserved for only those operators who had already passed the code test. Running Morse code practice on a local 2 meter repeater was one option, but nothing beats the excitement of practicing code on the worldwide bands and hooking up with another station thousands of miles away.

As of February 23, 2007, we can now take new General Class operators and introduce them to Morse code on the exciting worldwide bands!

Morse code is the ham radio operator's most basic language of short and long sounds, dits and dahs, or dots and dashes. Sailors have pounded SOS when trapped beneath a sailboat hull. In submarines, the tapping of Morse code gets the message through when there is no other way to communicate. Prisoners of war have tapped out Morse code messages, or BLINKED the code when being publicly displayed on television.

Ham operators use the code to get through when noise would otherwise cover up data or voice signals. Years ago, before road rage, fellow hams driving might greet each other by sending on their car horns H-I, a friendly salute to another ham.

So I encourage you to learn code. It is best mastered by sound along with memorizing the Morse code patterns seen on the upcoming pages. The pages show the number of dots and dashes to learn for a specific character, and learning the sound (rhythm) of Morse code is always the best way to practice. Let's see what these short sounds and long sounds are all about.

LOOKING AT MORSE CODE

The International Morse code, originally developed as the American Morse code by Samuel Morse, is truly international — all countries use it, and most commercial worldwide services employ operators who can recognize it. It is made up of short and long duration sounds. Long sounds, called "dahs," are three times longer than short sounds, called "dits." *Figure 2-1* shows the time intervals for Morse code sounds and spaces. *Figure 2-3,* on the next page, indicates the sounds for all the CW characters and symbols.

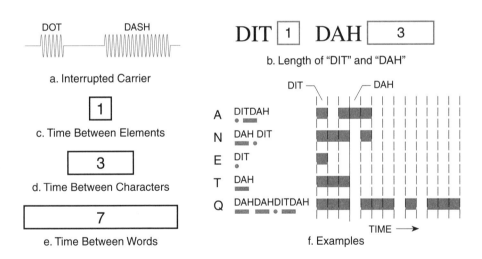

Figure 2-1. Time Intervals for Morse Code

a. Alphabet

LETTER	Composed of:	Sounds like:	LETTER	Composed of:	Sounds like:
A	• —	didah	N	— •	dahdit
B	— • • •	dahdididit	O	— — —	dahdahdah
C	— • — •	dahdidahdit	P	• — — •	didahdahdit
D	— • •	dahdidit	Q	— — • —	dahdahdidah
E	•	dit	R	• — •	didahdit
F	• • — •	dididahdit	S	• • •	dididit
G	— — •	dahdahdit	T	—	dah
H	• • • •	didididit	U	• • —	dididah
I	• •	didit	V	• • • —	didididah
J	• — — —	didahdahdah	W	• — —	ditdahdah
K	— • —	dahdidah	X	— • • —	dahdididah
L	• — • •	didahdidit	Y	— • — —	dahdidahdah
M	— —	dahdah	Z	— — • •	dahdahdidit

b. Special Signals and Punctuation

CHARACTER	Meaning:	Composed of:	Sounds like:
A̅R̅	(end of message)	• — • — •	didahdidahdit
K	invitation to transmit (go ahead)	— • —	dahdidah
S̅K̅	End of work	• • • — • —	didididahdidah
S O S	International distress call	• • • — — — • • •	didididahdahdahdididit
V	Test letter (V)	• • • —	didididah
R	Received, OK	• — •	didahdit
B̅T̅	Break or Pause	— • • • —	dahdidididah
D̅N̅	Slant Bar	— • • — •	dahdididahdit
K̅N̅	Back to You Only	— • — — •	dahdidahdahdit
Period		• — • — • —	didahdidahdidah
Comma		— — • • — —	dahdahdididahdah
Question mark		• • — — • •	dididahdahdidit
@	For Web Address	• — — • — •	didahdahdidahdi

c. Numerals

NUMBER	Composed of:	Sounds like:
1	• — — — —	didahdahdahdah
2	• • — — —	dididahdahdah
3	• • • — —	didididahdah
4	• • • • —	didididdah
5	• • • • •	dididididit
6	— • • • •	dahdidididit
7	— — • • •	dahdahdididit
8	— — — • •	dahdahdahdidit
9	— — — — •	dahdahdahdahdit
Ø	— — — — —	dahdahdahdahdah

Figure 2-3. Morse Code and Its Sound

CODE KEY

Morse code is usually sent by using a code key. A typical one is shown
Figure 2-2a. Normally it is mounted on a thin piece of wood or plexigl
sure that what you mount it on is thin; if the key is raised too high, it wi
uncomfortable to the wrist. The correct sending position for the hand is shown in
Figure 2-2b.

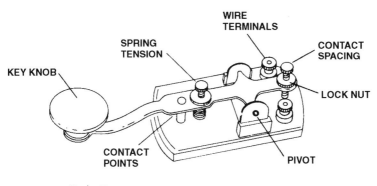

a. Code Key

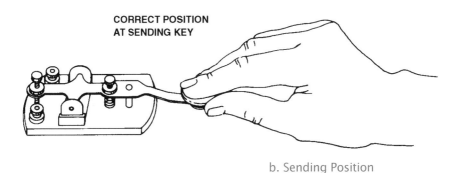

b. Sending Position

Figure 2-2. Code Key for Sending Code

LEARNING MORSE CODE

The reason you are learning the Morse code is to be able to operate all modes on the worldwide bands—including CW. Here are five suggestions (four serious ones) on how to learn the code:

1. Memorize the code from the code charts in this book.
2. Use my fun audio course available at all ham radio stores, and from the W5YI Group.
3. Go out and spend $1,000 and buy a worldwide radio, and listen to the code live and on the air. You don't need to spend that much, but you can listen to Morse code practice on the air, as shown in *Table 2-1*.
4. Use a code key and oscillator to practice sending the code. Believe it or not, someday you're actually going to do code over the live airwaves, using this same code key hooked up to your new megabuck transceiver.
5. Play with code programs on your computer, and *have fun!*

Table 2-1. Radio Frequencies and Times for Code Reception

Pacific	Mountain	Central	Eastern	Mon.	Tue.	Wed.	Thu.	Fri.
6 a.m.	7 a.m.	8 a.m.	9 a.m.		Fast Code	Slow Code	Fast Code	Slow Code
7 a.m. – 1 p.m.	8 a.m. – 2 p.m.	9 a.m. – 3 p.m.	10 a.m. – 4 p.m.	**VISITING OPERATOR TIME**				
1 p.m.	2 p.m.	3 p.m.	4 p.m.	Fast Code	Slow Code	Fast Code	Slow Code	Fast Code
2 p.m.	3 p.m.	4 p.m.	5 p.m.	Code Bulletin				
3 p.m.	4 p.m.	5 p.m.	6 p.m.	Digital Bulletin				
4 p.m.	5 p.m.	6 p.m.	7 p.m.	Slow Code	Fast Code	Slow Code	Fast Code	Slow Code
5 p.m.	6 p.m.	7 p.m.	8 p.m.	Code Bulletin				
6 p.m.	7 p.m.	8 p.m.	9 p.m.	Digital Bulletin				
6:45 p.m.	7:45 p.m.	8:45 p.m.	9:45 p.m.	Voice Bulletin				
7 p.m.	8 p.m.	9 p.m.	10 p.m.	Fast Code	Slow Code	Fast Code	Slow Code	Fast Code
8 p.m.	9 p.m.	10 p.m.	11 p.m.	Code Bulletin				

CW is broadcast on the following MHz frequencies: 1.8025, 3.5815, 7.0475, 14.0475, 18.0975, 21.0675, 28.0675, and 147.555. W1AW schedule courtesy of *QST* magazine.

DATA broadcasts: 3.5975, 7.095, 14.095, 18.1025, 21.095, 28.095 and 147.555 MHz

VOICE broadcasts: 1.855, 3.990, 7.290, 14.290, 18.160, 21.390, 28.590 and 147.555 MHz

CODE COURSES ON CDs AND CASSETTE TAPES

Five words per minute is so slow, and so easy, that many ham radio applicants learn it completely in a single week! You can do it, too, by using the code CDs and tapes mentioned above.

Code courses personally recorded by me make code learning *fun*. They will train you to send and receive the International Morse code in just a few short weeks. They are narrated and parallel the instructions in this book. The CDs have code characters generated at a 15-wpm character rate, spaced out to a 5-wpm word rate. This is known as Farnsworth spacing.

Getting Started

The hardest part of learning the code is taking the first CD out of the case, putting it in your player, and pushing the play button! Try it, and you will be over your biggest hurdle. After that, the CDs will talk you through the code in no time at all.

The CDs make code learning *fun*. You'll hear how humor has been added to the learning process to keep your interest high. Since ham radio is a hobby, there's no reason we can't poke ourselves in the ribs and have a little fun learning the code as part of this hobby experience. Okay, you're still not convinced — you probably have already made up your mind that trying to learn the code will be the hardest part of being a ham. It will not. Give yourself a fair chance. Don't get discouraged. Have patience and remember these important reminders when practicing to learn the Morse code:

- Learn the code by sound. Don't stare at the tiny dots and dashes that we have here in the book — the dit and dah sounds on the CDs and on the air and with your practice keyer will ultimately create an instant letter at your fingertips and into the pencil.
- *Never* scribble down dots or dashes if you forget a letter. Just put a small dash on your paper for a missed letter. You can go back and figure out what the word is by the letters you did copy!
- Practice only with fast code characters; 15-wpm character speed, spaced down to 5-wpm speed, is ideal.
- Practice the code by writing it down whenever possible. This further trains your brain and hand to work together in a subconscious response to the sounds you hear. (Remember Pavlov and his dog "Spot"?)
- Practice only for 15 minutes at a time. The CDs will tell you when to start and when to stop. Your brain and hand will lose that sharp edge once you go beyond 16 minutes of continuous code copy. You will learn much faster with five 15-minute practices per day than a one-hour marathon at night.
- Stay on course with the cassette instructions. Learn the letters, numbers, punctuation marks, and operating signals in the order they are presented. My code teaching system parallels that of the American Radio Relay League, Boy Scouts of America, the Armed Forces, and has worked for thousands in actual classroom instruction.

It was no accident that Samuel Morse gave the single dit for the letter "E" which occurs most often in the English language. He determined the most used letters in the alphabet by counting letters in a printer's type case. He reasoned a printer would have more of the most commonly-used letters. It worked! With just the first lesson, you will be creating simple words and simple sentences with no previous background.

Table 2-2 shows the sequence of letters, punctuation marks, operating signals, and numbers covered in six lessons on the CDs recorded by me.

Table 2-2. Sequence of Lessons on Cassettes	
▪ Lesson 1	E T M A N I S O S̄K̄ Period
▪ Lesson 2	R U D C 5 Ø A̅R̅ Question Mark
▪ Lesson 3	K P B G W F H B̅T̅ Comma
▪ Lesson 4	Q L Y J X V Z D̅N̅ 1 2 3 4 6 7 8 9
▪ Lesson 5	Random code with narrated answers
▪ Lesson 6	A typical 5-wpm code test

CODE KEY AND OSCILLATOR – HAM RECEIVER

All worldwide ham transceivers have provisions for a code key to be plugged in for both CW practice off the air as well as CW operating on the air. If you already own a worldwide set, chances are all you will need is a code key for some additional code-sending practice.

Read over your worldwide radio instruction manual where it talks about hooking up the code key. For code practice, read the notes about operating with a "side tone" but not actually going on the air. This "side tone" capability of most worldwide radios will eliminate your need for a separate code oscillator.

Code Key and Oscillator — Separate Unit

Many students may wish to simply buy a complete code key and oscillator set. They are available from local electronic outlets or through advertisements in the ham magazines.

Look again at the code key in *Figure 2-2a.* Note the terminals for the wires. Connect wires to these terminals and tighten the terminals so the wires won't come loose. The two wires will go either to a code oscillator set or to a plug that connects into your ham transceiver. Hook up the wires to the plug as described in your ham transceiver instruction book or the code oscillator set instruction book.

Mount the key firmly, as previously described, then adjust the gap between the contact points. With most new telegraph keys, you will need a pair of pliers to loosen the contact adjustment knob. It's located on the very end of your keyer. First loosen the lock nut, then screw down the adjustment until you get a gap no wider than the thickness of a business card. You want as little space as possible between the points. The contact points are located close to the sending plastic knob.

Now turn on your set or oscillator and listen. If your hear a constant tone, check that the right-hand movable shorting bar is not closed. If it is, swing it open. Adjust the spring tension adjustment screw so that you get a good "feel" each time you push down on the key knob. Adjust it tight enough to keep the contacts from closing while your fingers are resting on the key knob.

Pick up the key by the knob! This is the exact position your fingers should grasp the knob—one or two on top, and one or two on the side of it. Poking at the knob with one finger is unacceptable. Letting your fingers fly off the knob between dots and dashes (dits and dahs) also is not correct. As you are sending, you should be able to instantly pick up the whole key assembly to verify proper finger position.

Your arm and wrist should barely move as you send CW. All the action is in your hand — and it should be almost effortless. Give it a try, and look at *Figure 2-2b* again to double-check your hand position.

Letting someone else use the key to send CW to you will also help you learn the code.

Morse Code Computer Software

The newest way to learn Morse code is through computer-aided instruction. There are many good PC programs on the market that not only teach you the characters, but build speed and allow you to take actual telegraphy examinations, which the computer constructs. Personal computer programs also can be used to make audio tapes on your tape recorder so you can listen to them on the cassette player in your car.

A big advantage of computer-aided Morse code learning is that you can easily customize the program to fit your own needs! You can select the sending speed, Farnsworth character-spacing speed, duration of transmission, number of characters in a random group, tone frequency — and more!

Some have built-in "weighting." That means the software will determine your weaknesses and automatically adjust future sending to give you more study on your problem characters! All Morse code software programs transmit the tone by keying the PC's internal speaker. Some generate a clearer audio tone through the use of external oscillators or internal computer sound cards.

Here are three web addresses of sites that offer Morse code instruction for your computer:

http://lcwo.net
http://numorse.com/
http://www.dxzone.com/catalog/operatingmodes/morsecode/
    learningmorsecode/

Morse Code Audio Courses

Are you an audio learner? I've recorded 3 audio courses for learning Morse Code. My 8-CD code course teaches CW at a beginner level of 0 to 5 words per minute. To increase your speed, I have 2 CW cassette tape sets that will help you increase your speed to 13 wpm, and then all the way up to 20 wpm. And, if you knew CW from the Scouts or the Service, I have a 2-CD refresher course to get you back into sending and receiving the dits and dahs.

Start listening to my voice, and see how easy it is to master the dots and dashes. Continuously push yourself to learn the letters that contain more dots and dashes in them, and work those CDs and tapes regularly. Keep your copy in a spiral-bound notebook to track your progress.

I hope to hear your CW call on the worldwide bands soon!

Need Gordo's Morse Code Courses?
Call the W5YI Group at 1-800-669-9594, or visit www.w5yi.org

U.S. VOLUNTEER EXAMINER COORDINATORS IN THE AMATEUR SERVICE

Anchorage ARC VEC
PO Box 670616
Chugiak, AK 99567-0616
907-338-0662
e-mail: jwiley@alaska.net
Internet: www.kl7aa.net/vec/vecmain.html

ARRL/VEC
225 Main Street
Newington, CT 06111-1494
860/594-0300
860/594-0339 (fax)
e-mail: vec@arrl.org
Internet: www.arrl.org

Central America VEC CAVEC Inc.
2751 Christian Lane NE
Huntsville, AL 35811
256-851-8896
e-mail: aa4ii@bellsouth.net
Internet: www.cavec@bellsouth.net

Golden Empire Amateur Radio Society
P.O. Box 508
Chico, CA 95927-0508
530/345-3515
e-mail: wa6zrt@sbcglobal.net

Greater Los Angeles Amateur Radio Group
9737 Noble Avenue
North Hills, CA 91343-2403
818/892-2068
e-mail: gla.arg@gte.net
Internet: www.glaarg.org

Jefferson Amateur Radio Club
P.O. Box 73665
Metairie, LA 70033
504-831-1613
e-mail: w5gad_vec@w5gad.org

Laurel Amateur Radio Club, Inc.
P.O. Box 1259
Laurel, MD 20725-1259
301/937-0394 (6-9 PM)
e-mail: aa3of@arrl.net
Internet: http://larcmdorg.doore.net/vec/

The Milwaukee Radio Amateurs Club, Inc.
P.O. Box 070695
Milwaukee, WI 53207-0695
262/797-6722
e-mail: tom@supremecom.biz
Internet: www.w9rh.org

MO-KAN VEC Coordinator
228 Tennessee Road
Richmond, KS 66080-9174
785-867-2011
e-mail: wo0e@lcwb.coop

SANDARC-VEC
P.O. Box 2446
La Mesa, CA 91943-2446
619/697-1475
e-mail: n6nyx@cox.net
Internet: www.sandarc.net

Sunnyvale VEC Amateur Radio Club, Inc.
P.O. Box 60307
Sunnyvale, CA 94088-0307
408/255-9000 (exam info 24 hours)
e-mail: vec@amateur-radio.org
Internet: www.amateur-radio.org

W4VEC
PO Box 482
China Grove, NC 28023-0482
336-841-7576
e-mail: raef@lexcominc.net
Internet: www.w4vec.com

Western Carolina Amateur Radio Society
VEC, Inc.
6702 Matterhorn Ct.
Knoxville, TN 37918-6314
865-687-5410
e-mail: mail@WcarsVec.org
Internet: www.wcarsvec.org

W5YI-VEC
POB 200065
Arlington, TX 76006-0065
817-274-0400
800-669-9594
e-mail: w5yi-vec@w5yi.org
Internet: www.w5yi-vec.org

THE W5YI RF SAFETY TABLES

(Developed by Fred Maia, W5YI, working in cooperation with the ARRL.)

There are two ways to determine whether your station's radio frequency signal radiation is within the MPE (Maximum Permissible Exposure) guidelines established by the FCC for **"controlled"** and **"uncontrolled"** environments. One way is direct **"measurement"** of the RF fields. The second way is through **"prediction"** using various antenna modeling, equations and calculation methods described in the FCC's *OET Bulletin 65* and *Supplement B.*

In general, most amateurs will not have access to the appropriate calibrated equipment to make precise field strength/power density measurements. The field-strength meters in common use by amateur operators are inexpensive, hand-held field strength meters that do not provide the accuracy necessary for reliable measurements, especially when different frequencies may be encountered at a given measurement location. It is more practical for amateurs to determine their PEP output power at the antenna and then look up the required distances to the controlled/uncontrolled environments using the following tables, which were developed using the prediction equations supplied by the FCC.

The FCC has determined that radio operators and their families are in the "controlled" environment and your neighbors and passers-by are in the "uncontrolled" environment. The estimated minimum compliance distances are in meters from the transmitting antenna to either the occupational/controlled exposure environment ("Con") or the general population/uncontrolled exposure environment ("Unc") using typical antenna gains for the amateur service and assuming 100% duty cycle and maximum surface reflection. Therefore, these charts represent the worst case scenario. They do not take into consideration compliance distance reductions that would be caused by:

(1) Feed line losses, which reduce power output at the antenna especially at the VHF and higher frequency levels.

(2) Duty cycle caused by the emission type. The emission type factor accounts for the fact that, for some modulated emission types that have a non-constant envelope, the PEP can be considerably larger than the average power. Multiply the distances by 0.4 if you are using CW Morse telegraphy, and by 0.2 for two-way SSB (single sideband) voice. There is no reduction for FM.

(3) Duty cycle caused by on/off time or "time-averaging." The RF safety guidelines permit RF exposures to be averaged over certain periods of time with the average not to exceed the limit for continuous exposure. The averaging time for occupational/controlled exposures is 6 minutes, while the averaging time for general population/uncontrolled exposures is 30 minutes. For example, if the relevant time interval for time-averaging is 6 minutes, an amateur could be exposed to two times the applicable power density limit for three minutes as long as he or she were not exposed at all for the preceding or following three minutes.

A routine evaluation is not required for vehicular mobile or hand-held transceiver stations. Amateur Radio operators should be aware, however, of the potential for exposure to RF electromagnetic fields from these stations, and take measures (such as reducing transmitting power to the minimum necessary, positioning the radiating antenna as far from humans as practical, and limiting continuous transmitting time) to protect themselves and the occupants of their vehicles.

Amateur Radio operators should also be aware that the FCC radio-frequency safety regulations address exposure to people — and not the strength of the signal. Amateurs may exceed the Maximum Permissible Exposure (MPE) limits as long as no one is exposed to the radiation.

How to read the chart: If you are radiating 500 watts from your 10 meter dipole (about a 3 dB gain), there must be at least 4.5 meters (about 15 feet) between you (and your family) and the antenna — and a distance of 10 meters (about 33 feet) between the antenna and your neighbors.

Medium and High Frequency Amateur Bands
All distances are in meters

Freq. (MF/HF) (MHz/Band)	Antenna Gain (dBi)	Peak Envelope Power (watts)							
		100 watts		500 watts		1000 watts		1500 watts	
		Con.	Unc.	Con.	Unc.	Con.	Unc.	Con.	Unc.
2.0 (160m)	0	0.1	0.2	0.3	0.5	0.5	0.7	0.6	0.8
2.0 (160m)	3	0.2	0.3	0.5	0.7	0.6	1.06	0.8	1.2
4.0 (75/80m)	0	0.2	0.4	0.4	1.0	0.6	1.3	0.7	1.6
4.0 (75/80m)	3	0.3	0.6	0.6	1.3	0.9	1.9	1.0	2.3
7.3 (40m)	0	0.3	0.8	0.8	1.7	1.1	2.5	1.3	3.0
7.3 (40m)	3	0.5	1.1	1.1	2.5	1.6	3.5	1.9	4.2
7.3 (40m)	6	0.7	1.5	1.5	3.5	2.2	4.9	2.7	6.0
10.15 (30m)	0	0.5	1.1	1.1	2.4	1.5	3.4	1.9	4.2
10.15 (30m)	3	0.7	1.5	1.5	3.4	2.2	4.8	2.6	5.9
10.15 (30m)	6	1.0	2.2	2.2	4.8	3.0	6.8	3.7	8.3
14.35 (20m)	0	0.7	1.5	1.5	3.4	2.2	4.8	2.6	5.9
14.35 (20m)	3	1.0	2.2	2.2	4.8	3.0	6.8	3.7	8.4
14.35 (20m)	6	1.4	3.0	3.0	6.8	4.3	9.6	5.3	11.8
14.35 (20m)	9	1.9	4.3	4.3	9.6	6.1	13.6	7.5	16.7
18.168 (17m)	0	0.9	1.9	1.9	4.3	2.7	6.1	3.3	7.5
18.168 (17m)	3	1.2	2.7	2.7	6.1	3.9	8.6	4.7	10.6
18.168 (17m)	6	1.7	3.9	3.9	8.6	5.5	12.2	6.7	14.9
18.168 (17m)	9	2.4	5.4	5.4	12.2	7.7	17.2	9.4	21.1
21.145 (15m)	0	1.0	2.3	2.3	5.1	3.2	7.2	4.0	8.8
21.145 (15m)	3	1.4	3.2	3.2	7.2	4.6	10.2	5.6	12.5
21.145 (15m)	6	2.0	4.6	4.6	10.2	6.4	14.4	7.9	17.6
21.145 (15m)	9	2.9	6.4	6.4	14.4	9.1	20.3	11.1	24.9
24.99 (12m)	0	1.2	2.7	2.7	5.9	3.8	8.4	4.6	10.3
24.99 (12m)	3	1.7	3.8	3.8	8.4	5.3	11.9	6.5	14.5
24.99 (12m)	6	2.4	5.3	5.3	11.9	7.5	16.8	9.2	20.5
24.99 (12m)	9	3.4	7.5	7.5	16.8	10.6	23.7	13.0	29.0
29.7 (10m)	0	1.4	3.2	3.2	7.1	4.5	10.0	5.5	12.2
29.7 (10m)	3	2.0	4.5	4.5	10.0	6.3	14.1	7.7	17.3
29.7 (10m)	6	2.8	6.3	6.3	14.1	8.9	19.9	10.9	24.4
29.7 (10m)	9	4.0	8.9	8.9	19.9	12.6	28.2	15.4	34.5

VHF/UHF Amateur Bands
All distances are in meters

| Freq. (MF/HF) (MHz/Band) | Antenna Gain (dBi) | Peak Envelope Power (watts) | | | | | | | |
| | | 50 watts | | 100 watts | | 500 watts | | 1000 watts | |
		Con.	Unc.	Con.	Unc.	Con.	Unc.	Con.	Unc.
50 (6m)	0	1.0	2.3	1.4	3.2	3.2	7.1	4.5	10.1
50 (6m)	3	1.4	3.2	2.0	4.5	4.5	10.1	6.4	14.3
50 (6m)	6	2.0	4.5	2.8	6.4	6.4	14.2	9.0	20.1
50 (6m)	9	2.8	6.4	4.0	9.0	9.0	20.1	12.7	28.4
50 (6m)	12	4.0	9.0	5.7	12.7	12.7	28.4	18.0	40.2
50 (6m)	15	5.7	12.7	8.0	18.0	18.0	40.2	25.4	56.8
144 (2m)	0	1.0	2.3	1.4	3.2	3.2	7.1	4.5	10.1
144 (2m)	3	1.4	3.2	2.0	4.5	4.5	10.1	6.4	14.3
144 (2m)	6	2.0	4.5	2.8	6.4	6.4	14.2	9.0	20.1
144 (2m)	9	2.8	6.4	4.0	9.0	9.0	20.1	12.7	28.4
144 (2m)	12	4.0	9.0	5.7	12.7	12.7	28.4	18.0	40.2
144 (2m)	15	5.7	12.7	8.0	18.0	18.0	40.2	25.4	56.8
144 (2m)	20	10.1	22.6	14.3	32.0	32.0	71.4	45.1	101.0
222 (1.25m)	0	1.0	2.3	1.4	3.2	3.2	7.1	4.5	10.1
222 (1.25m)	3	1.4	3.2	2.0	4.5	4.5	10.1	6.4	14.3
222 (1.25m)	6	2.0	4.5	2.8	6.4	6.4	14.2	9.0	20.1
222 (1.25m)	9	2.8	6.4	4.0	9.0	9.0	20.1	12.7	28.4
222 (1.25m)	12	4.0	9.0	5.7	12.7	12.7	28.4	18.0	40.2
222 (1.25m)	15	5.7	12.7	8.0	18.0	18.0	40.2	25.4	56.8
450 (70cm)	0	0.8	1.8	1.2	2.6	2.6	5.8	3.7	8.2
450 (70cm)	3	1.2	2.6	1.6	3.7	3.7	8.2	5.2	11.6
450 (70cm)	6	1.6	3.7	2.3	5.2	5.2	11.6	7.4	16.4
450 (70cm)	9	2.3	5.2	3.3	7.3	7.3	16.4	10.4	23.2
450 (70cm)	12	3.3	7.3	4.6	10.4	10.4	23.2	14.7	32.8
902 (33cm)	0	0.6	1.3	0.8	1.8	1.8	4.1	2.6	5.8
902 (33cm)	3	0.8	1.8	1.2	2.6	2.6	5.8	3.7	8.2
902 (33cm)	6	1.2	2.6	1.6	3.7	3.7	8.2	5.2	11.6
902 (33cm)	9	1.6	3.7	2.3	5.2	5.2	11.6	7.3	16.4
902 (33cm)	12	2.3	5.2	3.3	7.3	7.3	16.4	10.4	23.2
1240 (23cm)	0	0.5	1.1	0.7	1.6	1.6	3.5	2.2	5.0
1240 (23cm)	3	0.7	1.6	1.0	2.2	2.2	5.0	3.1	7.0
1240 (23cm)	6	1.0	2.2	1.4	3.1	3.1	7.0	4.4	9.9
1240 (23cm)	9	1.4	3.1	2.0	4.4	4.4	9.9	6.3	14.0
1240 (23cm)	12	2.0	4.4	2.8	6.2	6.2	14.0	8.8	19.8

All distances are in meters. To convert from meters to feet multiply meters by 3.28. Distance indicated is shortest line-of-sight distance to point where MPE limit for appropriate exposure tier is predicted to occur.

AUTHORIZED FREQUENCY BANDS – AMATEUR SERVICE (for U.S. Amateur Stations operating from ITU-Region 2–North and South America)

METERS	Grandfathered² Novice	Technician / Technician	Tech. w/Code / Technician Plus	General / General	Advanced / Advanced	Extra Class / Extra Class
160				1800-2000 kHz/All	1800-2000 kHz/All	1800-2000 kHz/All
80/75	3525-3600 kHz/CW		3525-3600 kHz/CW	3525-3600 kHz/CW 3800-4000 kHz/Ph	3525-3600 kHz/CW 3700-4000 kHz/Ph	3500-4000 kHz/CW 3600-4000 kHz/Ph
40	7025-7125 kHz/CW		7025-7125 kHz/CW	7025-7125 kHz/CW 7175-7300 kHz/Ph	7025-7125 kHz/CW 7125-7300 kHz/Ph	7000-7300 kHz/CW 7125-7300 kHz/Ph
30				10.1-10.15 MHz/CW	10.1-10.15 MHz/CW	10.1-10.15 MHz/CW
20				14.025-14.15 MHz/CW 14.225-14.35 MHz/Ph	14.025-14.15 MHz/CW 14.175-14.35 MHz/Ph	14.0-14.35 MHz/CW 14.15-14.35 MHz/Ph
17				18.068-18.11 MHz/CW 18.11-18.168 MHz/Ph	18.068-18.11 MHz/CW 18.11-18.168 MHz/Ph	18.068-18.11 MHz/CW 18.11-18.168 MHz/Ph
15	21.025-21.2 MHz/CW		21.025-21.2 MHz/CW	21.025-21.2 MHz/CW 21.275-21.45 MHz/Ph	21.025-21.2 MHz/CW 21.225-21.45 MHz/Ph	21.0-21.45 MHz/CW 21.2-21.45 MHz/Ph
12				24.89-24.99 MHz/CW 24.93-24.99 MHz/Ph	24.89-24.99 MHz/CW 24.93-24.99 MHz/Ph	24.89-24.99 MHz/CW 24.93-24.99 MHz/Ph
10	28.0-28.5 MHz/CW 28.3-28.5 MHz/Ph		28.0-28.5 MHz/CW 28.3-28.5 MHz/Ph	28.0-28.3 MHz/CW 28.3-29.7 MHz/Ph	28.0-28.3 MHz/CW 28.3-29.7 MHz/Ph	28.0-29.7 MHz/CW 28.3-29.7 MHz/Ph
6		50-54 MHz/CW 50.1-54 MHz/Ph		50-54 MHz/CW 50.1-54 MHz/Ph	50-54 MHz/CW 50.1-54 MHz/Ph	50-54 MHz/CW 50.1-54 MHz/Ph
2		144-148 MHz/CW 144.1-148 MHz/All	144-148 MHz/CW 144.1-148 MHz/All	144-148 MHz/CW 144.1-148 MHz/All	144-148 MHz/CW 144.1-148 MHz/All	144-148 MHz/CW 144.1-148 MHz/All
1.25	222-225 MHz/All	³222-225 MHz/All	222-225 MHz/All	222-225 MHz/All	222-225 MHz/All	222-225 MHz/All
0.70		420-450 MHz/All	420-450 MHz/All	420-450 MHz/All	420-450 MHz/All	420-450 MHz/All
0.33		902-928 MHz/All	902-928 MHz/All	902-928 MHz/All	902-928 MHz/All	902-928 MHz/All
0.23	1270-1295 MHz/All	1240-1300 MHz/All	1240-1300 MHz/All	1240-1300 MHz/All	1240-1300 MHz/All	1240-1300 MHz/All

1 Effective 4-15-00 2 Prior to 4-15-00 3 Effective 2/1/94 219-220 MHz is authorized for point-to-point fixed digital message forwarding systems.

Note: Morse code (CW, A1A) may be used on any frequency allocated to the amateur service. Telephony emission (abbreviated Ph above) authorized on certain bands as indicated. Higher class licensees may use slow-scan television and facsimile emissions on the Phone bands; radio teletype/digital on the CW bands. All amateur modes and emissions are authorized above 144.1 MHz. In actual practice, the modes/emissions used are somewhat more complicated than shown above due to the existence of various band plans and "gentlemen's agreements" concerning where certain operations should take place.

The following CEPT countries allow U.S. Amateurs to operate in their countries without a reciprocal license. Be sure to carry a copy of your FCC license and FCC Public Notice DA99-1098.

Austria	France & its	Luxembourg	Sweden
Belgium	possessions	Monaco	Switzerland
Bosnia & Herzegovina	Germany	Montenegro	Turkey
Bulgaria	Greenland	Netherlands	United Kingdom & its
Croatia	Hungary	Netherlands Antilles	possessions
Cyprus	Iceland	Norway	
Czech Republic	Ireland	Portugal	
Denmark	Italy	Romania	
Estonia	Latvia	Slovak Republic	
Faroe Islands	Liechtenstein	Slovenia	
Finland	Lithuania	Spain	

List of Countries Permitting Third-Party Traffic

Country	Call Sign Prefix	Country	Call Sign Prefix	Country	Call Sign Prefix
Antigua and Barbuda	V2	El Salvador	YS	Paraguay	ZP
Argentina	LU	The Gambia	C5	Peru	OA
Australia	VK	Ghana	9G	Philippines	DU
Austria, Vienna	4U1VIC	Grenada	J3	St. Christopher & Nevis	V4
Belize	V3	Guatemala	TG	St. Lucia	J6
Bolivia	CP	Guyana	8R	St. Vincent & Grenadines. .	J8
Bosnia-Herzegovina	T9	Haiti	HH	Sierra Leone	9L
Brazil	PY	Honduras	HR	South Africa	ZS
Canada	VE, VO, VY	Israel	4X	Swaziland	3D6
Chile	CE	Jamaica	6Y	Trinidad and Tobago	9Y
Colombia	HK	Jordan	JY	Turkey	TA
Comoros	D6	Liberia	EL	United Kingdom	GB*
Costa Rica	TI	Marshall Is	V6	Uruguay	CX
Cuba	CO	Mexico	XE	Venezuela	YV
Dominica	J7	Micronesia	V6	ITU-Geneva	4U1ITU
Dominican Republic	HI	Nicaragua	YN	VIC-Vienna	4U1VIC
Ecuador	HC	Panama	HP		

Countries Holding U.S. Reciprocal Agreements

Antigua, Barbuda	Chile	Greece	Liberia	Seychelles
Argentina	Colombia	Greenland	Luxembourg	Sierra Leone
Australia	Costa Rica	Grenada	Macedonia	Solomon Islands
Austria	Croatia	Guatemala	Marshall Is.	South Africa
Bahamas	Cyprus	Guyana	Mexico	Spain
Barbados	Denmark	Haiti	Micronesia	St. Lucia
Belgium	Dominica	Honduras	Monaco	St. Vincent and
Belize	Dominican Rep.	Iceland	Netherlands	Grenadines
Bolivia	Ecuador	India	Netherlands Ant.	Surinam
Bosnia-	El Salvador	Indonesia	New Zealand	Sweden
Herzegovina	Fiji	Ireland	Nicaragua	Switzerland
Botswana	Finland	Israel	Norway	Thailand
Brazil	France[2]	Italy	Panama	Trinidad, Tobago
Canada[1]	Germany	Jamaica	Paraguay	Turkey
		Japan	Papua New Guinea	Tuvalu
1. Do not need reciprocal permit		Jordan	Peru	United Kingdom[3]
2. Includes all French Territories		Kiribati	Philippines	Uruguay
3. Includes all British Territories		Kuwait	Portugal	Venezuela

QUESTION POOL SYLLABUS

The syllabus used by the NCVEC Question Pool Committee to develop the question pool is included here as an aid in studying the subelements and topic groups. Reviewing the syllabus will give you an understanding of how the question pool is used to develop the Element 3 General Class written examination. Remember, one question will be taken from each topic group within each subelement to create your exam.

2011-15 Element 3 General Class Syllabus

G1 – Commission's Rules
[5 Exam Questions – 5 Groups]
G1A – General Class control operator frequency privileges; primary and secondary allocations
G1B – Antenna structure limitations; good engineering and good amateur practice; beacon operation; restricted operation; retransmitting radio signals
G1C – Transmitter power regulations; data emission standards
G1D – Volunteer Examiners and Volunteer Examiner Coordinators; temporary identification
G1E – Control categories; repeater regulations; harmful interference; third party rules; ITU regions

G2 – Operating Procedures
[5 Exam Questions – 5 Groups]
G2A – Phone operating procedures; USB/LSB utilization conventions; procedural signals; breaking into a QSO in progress; VOX operation
G2B – Operating courtesy; band plans, emergencies, including drills and emergency communications
G2C – CW operating procedures and procedural signals, Q signals and common abbreviations; full break in
G2D – Amateur Auxiliary; minimizing interference; HF operations
G2E – Digital operating: procedures, procedural signals and common abbreviations

G3 – Radio Wave Propagation
[3 Exam Questions – 3 Groups]
G3A – Sunspots and solar radiation; ionospheric disturbances; propagation forecasting and indices
G3B – Maximum Usable Frequency; Lowest Usable Frequency; propagation
G3C – Ionospheric layers; critical angle and frequency; HF scatter; Near Vertical Incidence Sky waves

G4 – Amateur Radio Practices
[5 Exam Questions – 5 groups]
G4A -Station Operation and setup
G4B – Test and monitoring equipment; two-tone test
G4C – Interference with consumer electronics; grounding; DSP
G4D – Speech processors; S meters; sideband operation near band edges
G4E – HF mobile radio installations; emergency and battery powered operation

G5 – Electrical Principles
[3 exam questions – 3 groups]
G5A – Reactance; inductance; capacitance; impedance; impedance matching
G5B – The Decibel; current and voltage dividers; electrical power calculations; sine wave root-mean-square (RMS) values; PEP calculations
G5C – Resistors, capacitors and inductors in series and parallel; transformers

G6 – Circuit Components
[3 exam question – 3 groups]
G6A – Resistors; capacitors; inductors
G6B – Rectifiers; solid state diodes and transistors; vacuum tubes; batteries
G6C – Analog and digital integrated circuits (IC's); microprocessors; memory; I/O devices; microwave IC's (MMIC's); display devices

G7 – Practical Circuits
[3 exam question – 3 groups]
G7A – Power supplies; schematic symbols
G7B – Digital circuits; amplifiers and oscillators
G7C – Receivers and transmitters; filters, oscillators

G8 – Signals & Emissions
[2 exam questions – 2 groups]
G8A – Carriers and modulation: AM; FM; single and double sideband; modulation envelope; overmodulation
G8B – Frequency mixing; multiplication; HF data communications; bandwidths of various modes; deviation

G9 – Antennas & Feed Lines
[4 exam questions – 4 groups]
G9A – Antenna feed lines: characteristic impedance and attenuation; SWR calculation, measurement and effects; matching networks
G9B – Basic antennas
G9C – Directional antennas
G9D – Specialized antennas

G0 – Electrical & RF Safety
[2 exam Questions – 2 groups]
G0A – RF safety principles, rules and guidelines; routine station evaluation
G0B – Safety in the ham shack: electrical shock and treatment, safety grounding, fusing, interlocks, wiring, antenna and tower safety

2007-11 ELEMENT 3 Q&A CROSS REFERENCE

The following cross reference presents all 486 question numbers in numerical order included in the 2011-15 Element 3 General Class question pool, followed by the page number on which the question begins in the book. This will allow you to locate specific questions by question number. Note: one question was deleted from the pool by the Question Pool Committee, resulting in an active pool of 456 questions. The deleted questions do not appear in the book.

Question	Page	Question	Page	Question	Page	Question	Page	Question	Page
G1 – Commission's Rules		G1E01	39	G2D05	39	G3C01	79	G4C07	148
G1A01	26	G1E02	27	G2D06	83	G3C02	76	G4C08	150
G1A02	28	G1E03	29	G2D07	30	G3C03	75	G4C09	149
G1A03	29	G1E04	49	G2D08	34	G3C04	76	G4C10	149
G1A04	30	G1E05	40	G2D09	34	G3C05	80	G4C11	105
G1A05	29	G1E06	50	G2D10	49	G3C06	81	G4C12	105
G1A06	27	G1E07	40	G2D11	156	G3C07	82	G4C13	107
G1A07	31	G1E08	41			G3C08	81		
G1A08	28	G1E09	25	G2E01	63	G3C09	81	G4D01	91
G1A09	31	G1E10	50	G2E02	63	G3C10	82	G4D02	92
G1A10	28			G2E03	66	G3C11	153	G4D03	93
G1A11	27	**G2 – Operating**		G2E04	62	G3C12	79	G4D04	107
G1A12	32	**Procedures**		G2E05	62	G3C13	88	G4D05	100
G1A13	32	G2A01	52	G2E06	62			G4D06	107
G1A14	31	G2A02	53	G2E07	62	**G4 – Amateur Radio**		G4D07	100
G1A15	30	G2A03	52	G2E08	66	**Practices**		G4D08	48
		G2A04	52	G2E09	63	G4A01	106	G4D09	48
G1B01	177	G2A05	51	G2E10	64	G4A02	60	G4D10	48
G1B02	60	G2A06	51	G2E11	64	G4A03	47	G4D11	48
G1B03	59	G2A07	52	G2E12	68	G4A04	102		
G1B04	72	G2A08	46	G2E13	68	G4A05	92	G4E01	167
G1B05	37	G2A09	53			G4A06	162	G4E02	166
G1B06	37	G2A10	49	**G3 – Radio Wave**		G4A07	93	G4E03	130
G1B07	38	G2A11	46	**Propagation**		G4A08	102	G4E04	131
G1B08	36			G3A01	83	G4A09	95	G4E05	166
G1B09	38	G2B01	46	G3A02	86	G4A10	61	G4E06	166
G1B10	59	G2B02	71	G3A03	86	G4A11	48	G4E07	151
G1B11	36	G2B03	47	G3A04	84	G4A12	47	G4E08	129
G1B12	36	G2B04	54	G3A05	84	G4A13	106	G4E09	129
		G2B05	47	G3A06	86	G4A14	64	G4E10	130
G1C01	29	G2B06	45	G3A07	88			G4E11	130
G1C02	28	G2B07	45	G3A08	86	G4B01	94		
G1C03	30	G2B08	32	G3A09	84	G4B02	95	**G5 – Electrical**	
G1C04	28	G2B09	72	G3A10	84	G4B03	95	**Principles**	
G1C05	27	G2B10	73	G3A11	83	G4B04	95	G5A01	142
G1C06	31	G2B11	71	G3A12	85	G4B05	128	G5A02	140
G1C07	65	G2B12	71	G3A13	85	G4B06	127	G5A03	140
G1C08	65			G3A14	87	G4B07	169	G5A04	141
G1C09	66	G2C01	59	G3A15	87	G4B08	168	G5A05	142
G1C10	65	G2C02	57	G3A16	86	G4B09	168	G5A06	141
G1C11	66	G2C03	57			G4B10	173	G5A07	143
		G2C04	59			G4B11	167	G5A08	143
		G2C05	54			G4B12	168	G5A09	140
G1D01	24	G2C06	54	G3B01	83	G4B13	167	G5A10	142
G1D02	42	G2C07	55	G3B02	78	G4B14	127	G5A11	142
G1D03	24	G2C08	59	G3B03	78	G4B15	94	G5A12	143
G1D04	43	G2C09	57	G3B04	77	G4B16	94	G5A13	143
G1D05	42	G2C10	57	G3B05	77				
G1D06	24	G2C11	57	G3B06	80	G4C01	150	G5B01	100
G1D07	42			G3B07	80	G4C02	151	G5B02	138
G1D08	43	G2D01	35	G3B08	77	G4C03	151	G5B03	122
G1D09	23	G2D02	35	G3B09	76	G4C04	151	G5B04	122
G1D10	43	G2D03	35	G3B10	78	G4C05	148	G5B05	122
		G2D04	82	G3B11	80	G4C06	150	G5B06	98
				G3B12	78				

Glossary

Amateur communication: Noncommercial radio communication by or among amateur stations solely with a personal aim and without personal or business interest.

Amateur operator/primary station license: An instrument of authorization issued by the FCC comprised of a station license, and also incorporating an operator license indicating the class of privileges.

Amateur operator: A person holding a valid license to operate an amateur station issued by the FCC. Amateur operators are frequently referred to as ham operators.

Amateur Radio services: The amateur service, the amateur-satellite service, and the radio amateur civil emergency service.

Amateur-satellite service: A radiocommunication service using stations on Earth satellites for the same purpose as those of the amateur service.

Amateur service: A radiocommunication service for the purpose of self-training, intercommunication and technical investigations carried out by amateurs; that is, duly authorized persons interested in radio technique solely with a personal aim and without pecuniary interest.

Amateur station: A station licensed in the amateur service embracing necessary apparatus at a particular location used for amateur communication.

AMSAT: Radio Amateur Satellite Corporation, a nonprofit scientific organization. (P.O. Box #27, Washington, DC 20044)

ANSI: American National Standards Institute. A non-government organization that develops recommended standards for a variety of applications.

APRS: Automatic Position Radio System, which takes GPS (Global Positioning System) information and translates it into an automatic packet of digital information.

ARES: Amateur Radio Emergency Service — the emergency division of the American Radio Relay League. Also see RACES

ARRL: American Radio Relay League, national organization of U.S. Amateur Radio operators. (225 Main Street, Newington, CT 06111)

Audio Frequency (AF): The range of frequencies that can be heard by the human ear, generally 20 hertz to 20 kilohertz.

Automatic control: The use of devices and procedures for station control without the control operator being present at the control point when the station is transmitting.

Automatic Volume Control (AVC): A circuit that continually maintains a constant audio output volume in spite of deviations in input signal strength.

Beam or Yagi antenna: An antenna array that receives or transmits RF energy in a particular direction. Usually rotatable.

Block diagram: A simplified outline of an electronic system where circuits or components are shown as boxes.

Broadcasting: Information or programming transmitted by radio means intended for the general public.

Bulletin No. 65: The Office of Engineering & Technology bulletin that provides specified safety guidelines for human exposure to radiofrequency (RF) radiation.

Business communications: Any transmission or communication the purpose of which is to facilitate the regular business or commercial affairs of any party. Business communications are prohibited in the amateur service.

Call Book: A published list of all licensed amateur operators available in North American and Foreign editions.

Call sign: The FCC systematically assigns each amateur station its primary call sign.

Certificate of Successful Completion of Examination (CSCE): A certificate providing examination credit for 365 days. Both written and code credit can be authorized.

Coaxial cable, Coax: A concentric, two-conductor cable in which one conductor surrounds the other, separated by an insulator.

Controlled Environment: Involves people who are aware of and who can exercise control over radiofrequency exposure. Controlled exposure limits apply to both occupational workers and Amateur Radio operators and their immediate households.

Control operator: An amateur operator designated by the licensee of an amateur station to be responsible for the station transmissions.

Coordinated repeater station: An amateur repeater station for which the transmitting and receiving frequencies have been recommended by the recognized repeater coordinator.

Coordinated Universal Time (UTC): (Also Greenwich Mean Time, UCT or Zulu time.) The time at the zero-degree (0°) Meridian which passes through Greenwich, England. A universal time among all amateur operators.

Crystal: A quartz or similar material which has been ground to produce natural vibrations of a specific frequency. Quartz crystals produce a high degree of frequency stability in radio transmitters.

CW: See Morse code.

Dipole antenna: The most common wire antenna. Length is equal to one-half of the wavelength. Fed by coaxial cable.

Dummy antenna: A device or resistor which serves as a transmitter's antenna without radiating radio waves. Generally used to tune up a radio transmitter.

Duplexer: A device that allows a single antenna to be simultaneously used for both reception and transmission.

Duty cycle: As applies to RF safety, the percentage of time that a transmitter is "on" versus "off" in a 6- or 30-minute time period.

Effective Radiated Power (ERP): The product of the transmitter (peak envelope) power, expressed in watts, delivered to the antenna, and the relative gain of an antenna over that of a half-wave dipole antenna.

Electromagnetic radiation: The propagation of radiant energy, including infrared, visible light, ultraviolet, radiofrequency, gamma and X-rays, through space and matter.

Emergency communication: Any amateur communication directly relating to the immediate safety of life of individuals or the immediate protection of property.

Examination Element: The written theory exam or CW test required for various classes of FCC Amateur Radio licenses. Technician must pass Element 2 written theory; General must pass Element 3 written theory plus Element 1 CW; Extra must pass Element 4 written theory.

Far Field: The electromagnetic field located at a great distance from a transmitting antenna. The far field begins at a distance that depends on many factors, including the wavelength and the size of the antenna. Radio signals are normally received in the far field.

FCC Form 605: The FCC application form used to apply for a new amateur operator/primary station license or to renew or modify an existing license.

Federal Communications Commission (FCC): A board of five Commissioners, appointed by the President, having the power to regulate wire and radio telecommunications in the U.S.

Feedline: A system of conductors that connects an antenna to a receiver or transmitter.

Field Day: Annual activity sponsored by the ARRL to demonstrate emergency preparedness of amateur operators.

Field strength: A measure of the intensity of an electric or magnetic field. Electric fields are measured in volts per meter; magnetic fields in amperes per meter.

Filter: A device used to block or reduce alternating currents or signals at certain frequencies while allowing others to pass unimpeded.

Frequency: The number of cycles of alternating current in one second.

Frequency coordinator: An individual or organization which recommends frequencies and other operating and/or technical parameters for amateur repeater operation in order to avoid or minimize potential interferences.

Frequency Modulation (FM): A method of varying a radio carrier wave by causing its frequency to vary in accordance with the information to be conveyed.

Frequency privileges: The transmitting frequency bands available to the various classes of amateur operators. The various Class privileges are listed in Part 97.301 of the FCC rules.

Ground: A connection, accidental or intentional, between a device or circuit and the earth or some common body and the earth or some common body serving as the earth.

Ground wave: A radio wave that is propagated near or at the earth's surface.

Handi-Ham system: Amateur organization dedicated to assisting handicapped amateur operators. (3915 Golden Valley Road, Golden Valley, MN 55422)

Harmful interference: Interference which seriously degrades, obstructs or repeatedly interrupts the operation of a radio communication service.

Harmonic: A radio wave that is a multiple of the fundamental frequency. The second harmonic is twice the fundamental frequency, the third harmonic, three times, etc.

Hertz: One complete alternating cycle per second. Named after Heinrich R. Hertz, a German physicist. The number of hertz is the frequency of the audio or radio wave.

High Frequency (HF): The band of frequencies that lie between 3 and 30 Megahertz. It is from these frequencies that radio waves are returned to earth from the ionosphere.

High-Pass filter: A device that allows passage of high frequency signals but attenuates the lower frequencies. When installed on a television set, a high-pass filter allows TV frequencies to pass while blocking lower-frequency amateur signals.

Inverse Square Law: The physical principle by which power density decreases as you get further away from a transmitting antenna. RF power density decreases by the inverse square of the distance.

Ionization: The process of adding or stripping away electrons from atoms or molecules. Ionization occurs when substances are heated at high temperatures or exposed to high voltages. It can lead to significant genetic damage in biological tissue.

Ionosphere: Outer limits of atmosphere from which HF amateur communications signals are returned to earth.

IRC: International Reply Coupon, a method of prepaying postage for a foreign amateur's QSL card.

Jamming: The intentional, malicious interference with another radio signal.

Key clicks, Chirps: Defective keying of a telegraphy signal sounding like tapping or high varying pitches.

Linear amplifier: A device that accurately reproduces a radio wave in magnified form.

Long wire: A horizontal wire antenna that is one wavelength or longer in length.

Low-Pass filter: Device connected to worldwide transmitters that inhibits passage of higher frequencies that cause television interference but does not affect amateur transmissions.

Machine: A ham slang word for an automatic repeater station.

Malicious interference: See jamming.

MARS: The Military Affiliate Radio System. An organization that coordinates the activities of amateur communications with military radio communications.

Maximum authorized transmitting power: Amateur stations must use no more than the maximum transmitter power necessary to carry out the desired communications. The maximum P.E.P. output power levels authorized Novices are 200 watts in the 80-, 40-, 15- and 10-meter bands, 25 watts in the 222-MHz band, and 5 watts in the 1270-MHz bands.

Maximum Permissible Exposure (MPE): The maximum amount of electric and magnetic RF energy to which a person may safely be exposed.

Maximum usable frequency (MFU): The highest

frequency that will be returned to earth from the ionosphere.

Medium frequency (MF): The band of frequencies that lies between 300 and 3,000 kHz (3 MHz).

Microwave: Electromagnetic waves with a frequency of 300 MHz to 300 GHz. Microwaves can cause heating of biological tissue.

Mobile operation: Radio communications conducted while in motion or during halts at unspecified locations.

Mode: Type of transmission such as voice, teletype, code, television, facsimile.

Modulate: To vary the amplitude, frequency, or phase of a radiofrequency wave in accordance with the information to be conveyed.

Morse code: The International Morse code, A1A emission. Interrupted continuous wave communications conducted using a dot-dash code for letters, numbers and operating procedure signs.

Near Field: The electromagnetic field located in the immediate vicinity of the antenna. Energy in the near field depends on the size of the antenna, its wavelength and transmission power.

Nonionizing radiation: Electromagnetic waves, or fields, which do not have the capability to alter the molecular structure of substances. RF energy is nonionizing radiation.

Novice operator: An FCC licensed, entry-level amateur operator in the amateur service.

Occupational exposure: See controlled environment.

OET: Office of Engineering & Technology, a branch of the FCC that has developed the guidelines for radiofrequency (RF) safety.

Ohm's law: The basic electrical law explaining the relationship between voltage, current and resistance. The current (I) in a circuit is equal to the voltage (E) divided by the resistance (R), or $I = E/R$.

OSCAR: "Orbiting Satellite Carrying Amateur Radio." A series of satellites designed and built by amateur operators of several nations.

Oscillator: A device for generating oscillations or vibrations of an audio or radiofrequency signal.

Packet radio: A digital method of communicating computer-to-computer. A terminal-node controller makes up the packet of data and directs it to another packet station.

Peak Envelope Power (PEP): 1. The power during one radiofrequency cycle at the crest of the modulation envelope, taken under normal operating conditions. 2. The maximum power that can be obtained from a transmitter.

Phone patch: Interconnection of amateur radio to the public switched telephone network, and operated by the control operator of the station.

Power density: A measure of the strength of an electro-magnetic field at a distance from its source. Usually expressed in milliwatts per square centimeter (mW/cm2). Far-field power density decreases according to the Law of Inverse Squares.

Power supply: A device or circuit that provides the appropriate voltage and current to another device or circuit.

Propagation: The travel of electromagnetic waves or sound waves through a medium.

Public exposure: See "uncontrolled" environment.

Q-signals: International three-letter abbreviations beginning with the letter Q used primarily to convey information using the Morse code.

QSL Bureau: An office that bulk processes QSL (radio confirmation) cards for (or from) foreign amateur operators as a postage-saving mechanism.

RACES (Radio Amateur Civil Emergency Service): A radio service using amateur stations for civil defense communications during periods of local, regional, or national emergencies.

Radiation: Electromagnetic energy, such as radio waves, traveling forth into space from a transmitter.

Radiofrequency (RF): The range of frequencies over 20 kilohertz that can be propagated through space.

Radiofrequency (RF) radiation: Electromagnetic fields or waves having a frequency between 3 kHz and 300 GHz.

Radiofrequency spectrum: The eight electromagnetic bands ranked according to their frequency and wavelength. Specifically, the very-low, low, medium, high, very-high, ultra-high, super-high, and extremely-high frequency bands.

Radio wave: A combination of electric and magnetic fields varying at a radiofrequency and traveling through space at the speed of light.

Repeater operation: Automatic amateur stations that retransmit the signals of other amateur stations.

Routine RF radiation evaluation: The process of determining if the RF energy from a transmitter exceeds the Maximum Permissible Exposure (MPE) limits in a controlled or uncontrolled environment.

RST Report: A telegraphy signal report system of Readability, Strength and Tone.

S-meter: A voltmeter calibrated from 0 to 9 that indicates the relative signal strength of an incoming signal at a radio receiver.

Selectivity: The ability of a circuit (or radio receiver) to separate the desired signal from those not wanted.

Sensitivity: The ability of a circuit (or radio receiver) to detect a specified input signal.

Short circuit: An unintended, low-resistance connection across a voltage source resulting in high current and possible damage.

Shortwave: The high frequencies that lie between 3 and 30 Megahertz that are propagated long distances.

Single-Sideband (SSB): A method of radio transmission in which the RF carrier and one of the sidebands is suppressed and all of the information is carried in the one remaining sideband.

Skip wave, Skip zone: A radio wave reflected back to earth. The distance between the radio transmitter and the site of a radio wave's return to earth.

Sky-wave: A radio wave that is refracted back to earth. Sometimes called an ionospheric wave.

Specific Absorption Rate (SAR): The time rate at which radiofrequency energy is absorbed into the human body.

Spectrum: A series of radiated energies arranged in order of wavelength. The radio spectrum extends from 20 kilohertz upward.

Spurious Emissions: Unwanted radiofrequency signals emitted from a transmitter that sometimes cause interference.

Station license, location: No transmitting station shall be operated in the amateur service without being licensed by the FCC. Each amateur station shall have one land location, the address of which appears in the station license.

Sunspot Cycle: An 11-year cycle of solar disturbances which greatly affects radio wave propagation.

Technician operator: An Amateur Radio operator who has successfully passed Element 2.

Technician-Plus: An amateur operator who has passed a 5-wpm code test in addition to Technician Class requirements.

Telegraphy: Communications transmission and reception using CW, International Morse code.

Telephony: Communications transmission and reception in the voice mode.

Telecommunications: The electrical conversion, switching, transmission and control of audio video and data signals by wire or radio.

Temporary operating authority: Authority to operate your amateur station while awaiting arrival of an upgraded license.

Terrestrial station location: Any location of a radio station on the surface of the earth including the sea.

Thermal effects: As applies to RF radiation, biological tissue damage resulting because of the body's inability to cope with or dissipate excessive heat.

Third-party traffic: Amateur communication by or under the supervision of the control operator at an amateur station to another amateur station on behalf of others.

Time-averaging: As applies to RF safety, the amount of electromagnetic radiation over a given time. The premise of time-averaging is that the human body can tolerate the thermal load caused by high, localized RF exposures for short periods of time.

Transceiver: A combination radio transmitter and receiver.

Transition region: Area where power density decreases inversely with distance from the antenna.

Transmatch: An antenna tuner used to match the impedance of the transmitter output to the transmission line of an antenna.

Transmitter: Equipment used to generate radio waves. Most commonly, this radio carrier signal is amplitude varied or frequency varied (modulated) with information and radiated into space.

Transmitter power: The average peak envelope power (output) present at the antenna terminals of the transmitter.

The term "transmitted" includes any external radio-frequency power amplifier which may be used.

Ultra High Frequency (UHF): Ultra high frequency radio waves that are in the range of 300 to 3,000 MHz.

Uncontrolled environment: Applies to those persons who have no control over their exposure to RF energy in the environment. Residences adjacent to ham radio installations are considered to be in an "uncontrolled" environment.

Upper Sideband (USB): The proper operating mode for sideband transmissions made in the new Novice 10-meter voice band. Amateurs generally operate USB at 20 meters and higher frequencies; lower sideband (LSB) at 40 meters and lower frequencies.

Very High Frequency (VHF): Very high frequency radio waves that are in the range of 30 to 300 MHz.

Volunteer Examiner: An amateur operator of at least a General Class level who prepares and administers amateur operator license examinations.

Volunteer Examiner Coordinator (VEC): A member of an organization which has entered into an agreement with the FCC to coordinate the efforts of volunteer examiners in preparing and administering examinations for amateur operator licenses.

Index

FREE
CQ Mini-Sub!

We'd like to introduce you to a Ham radio magazine that's fun to read, interesting from cover to cover and written so that you can understand it—**FREE!** The magazine is **CQ Amateur Radio**—read by thousands of people every month in 116 countries around the world. Get your FREE 3 issue mini-sub, courtesy **CQ** and **Master Publishing**.

CQ is aimed squarely at the **active** ham. You'll find features and columns covering the broad and varied landscape of the amateur radio hobby from contesting and DXing to satellites and the latest digital modes. Equipment reviews, projects and articles on the science as well as the art of radio communications—all in the pages of **CQ Amateur Radio**.

Reserve your FREE 3-month mini-sub today!

Remove this page, fill in your information below, fold, tape closed with the postage paid **CQ** address showing and mail today!

Send my FREE 3-issue CQ Mini-Sub to:

Name _____

Address _____

City _____ State _____ Zip _____

Make a great deal even better!

Add a one-year CQ subscription to your FREE mini-sub at a
**Special Introductory Rate—15 issues for only $29.95
a 60% savings off the newsstand price!**

Yes! I want to take advantage of this special 15-issue offer.

☐ Check/Money Order enclosed.
Bill my: ☐ Visa   ☐ MasterCard   ☐ AMEX   ☐ Discover
Insert your credit card number below:

Credit Card Expiration Date: _____

Fax your order to: 516-681-2926
Visit our web site: www.cq-amateur-radio.com

YI-G-07

The Worldwide Listening Guide

By John Figliozzi. Modeled on his popular *Worldwide Shortwave Listening Guide*, this new book explains radio listening in all today's formats – "live," on-demand, WiFi, podcast, terrestrial, satellite, internet, digital and, of course, analog, AM, FM and SW. Book includes a comprehensive programming guide to what can be heard how, when and where. Spiral-bound to open in a flat, easy-to-use format.   **WWLG $24.95**

Need to find a test location fast?
Need to change your address?
Need to renew your license?
Want a Vanity Call Sign?

Let the W5YI-VEC Help You!

The W5YI-VEC is a non-profit organization that coordinates Amateur Radio test sessions throughout the United States and in some foreign countries. Each Volunteer Examiner team is made up of 3 or more licensed Amateur Radio operators certified by The W5YI-VEC to administer ham radio license exams.

The W5YI-VEC can also help you with your FCC Amateur Radio license renewal, change of address, or obtaining a Vanity Call Sign.

Commercial Radio License Services

National Radio Examiners provides examination sites for FCC Commercial Radio Licenses, including GROL, MROP, and the GMDSS license. NRE also can assist you with your license renewal or change of address.

For help with any of these FCC Amateur or Commercial radio licensing services, call or visit:

1-800-669-9594
or visit www.w5yi.org